Drought, Flood, Fire

Every year, droughts, floods, and fires impact hundreds of millions of people and cause massive economic losses. Climate change is making these catastrophes more dangerous. Now. Not in the future: NOW. This book describes how and why climate change is already fomenting dire consequences, and will certainly make climate disasters worse in the near future. Chris Funk combines the latest science with compelling stories, providing a timely, accessible, and beautifully written synopsis of this critical topic. The book describes our unique and fragile Earth system, and the negative impacts humans are having on our support systems. It then examines recent disasters, including heat waves, extreme precipitation, hurricanes, fires, El Niños and La Niñas, and their human consequences. By clearly describing the dangerous impacts that are already occurring, Funk provides a clarion call for social change, yet also conveys the beauty and wonder of our planet, and hope for our collective future.

Chris Funk is an internationally renowned climate hazard scientist who develops data sets and forecasts that routinely help safeguard lives and livelihoods around the world. His publications focus on climate and climate change, highlighting how climate science can provide opportunities for predicting natural disasters, thus helping alleviate their dire consequences. His research has been featured in *Science* magazine, on TV, in mainstream print media, and on the radio. He currently directs the Climate Hazards Center at the University of California, Santa Barbara.

Drought, Flood, Fire
How Climate Change Contributes to Catastrophes

Chris Funk
University of California, Santa Barbara

CAMBRIDGE
UNIVERSITY PRESS

University Printing House, Cambridge CB2 8BS, United Kingdom

One Liberty Plaza, 20th Floor, New York, NY 10006, USA

477 Williamstown Road, Port Melbourne, VIC 3207, Australia

314–321, 3rd Floor, Plot 3, Splendor Forum, Jasola District Centre,
New Delhi – 110025, India

79 Anson Road, #06-04/06, Singapore 079906

Cambridge University Press is part of the University of Cambridge.

It furthers the University's mission by disseminating knowledge in the pursuit of education, learning, and research at the highest international levels of excellence.

www.cambridge.org
Information on this title: www.cambridge.org/9781108839877
DOI: 10.1017/9781108885348

© Chris Funk 2021

This publication is in copyright. Subject to statutory exception and to the provisions of relevant collective licensing agreements, no reproduction of any part may take place without the written permission of Cambridge University Press.

First published 2021

Printed in the United Kingdom by TJ Books Limited, Padstow Cornwall

A catalogue record for this publication is available from the British Library.

Library of Congress Cataloging-in-Publication Data
NAMES: Funk, Chris, author.
TITLE: Drought, flood, fire : how climate change contributes to catastrophes / Chris Funk, University of California, Santa Barbara.
DESCRIPTION: Cambridge, UK ; New York, NY : Cambridge University Press, 2021. | Includes bibliographical references and index.
IDENTIFIERS: LCCN 2020056635 (print) | LCCN 2020056636 (ebook) |
  ISBN 9781108839877 (hardback) | ISBN 9781108813792 (paperback) |
  ISBN 9781108885348 (epub)
SUBJECTS: LCSH: Climatic extremes. | Climatic changes.
CLASSIFICATION: LCC QC981.8.C53 F86 2021 (print) | LCC QC981.8.C53 (ebook) |
  DDC 363.34/1–dc23
LC record available at https://lccn.loc.gov/2020056635
LC ebook record available at https://lccn.loc.gov/2020056636

ISBN 978-1-108-83987-7 Hardback

Cambridge University Press has no responsibility for the persistence or accuracy of URLs for external or third-party internet websites referred to in this publication and does not guarantee that any content on such websites is, or will remain, accurate or appropriate.

CONTENTS

Acknowledgments   vii

1   Climate Extremes, Climate Attribution, Extreme Event Attribution   1

2   Welcome to an Awesome Planet: A Series of Delicate Balances Support Earth's Fragile Flame   21

3   The Earth Is a Negentropic System, or "the Bright Side of Empty"   40

4   Do-It-Yourself Climate Change Science   59

5   Temperature Extremes – Impacts and Attribution: Shocks, Exposure, and Vulnerability   92

6   Precipitation Extremes: Observations and Impacts   122

7   Hurricanes, Cyclones, and Typhoons   140

8   Conceptual Models of Climate Change and Prediction, and How They Relate to Floods and Fires   164

9   Climate Change Made the 2015–2016 El Niño More Extreme   186

10   Bigger La Niñas and the East African Climate Paradox   212

11   Fire and Drought in the Western United States   234

12 Fire and Australia's Black Summer   252

13 Driving toward +4°C on a Dixie® Cup Planet   268

14 We Can Afford to Wear a White Hat   287

*Appendix A Few Resources for Further Reading
          and Research*   305

Index   321

ACKNOWLEDGMENTS

First and foremost, I would like to thank the "first responders" from all walks of life and all over the world who inspired me to write this book. Every day, you put yourselves on the line to help others, teaching us what it means to be "careful" human beings. This book grew out of dozens of conversations with volunteer firefighters here in California (Mike Williams, Ted Adams, Rocky Siegel, and Yesi Thomas) and humanitarian relief workers focused on Africa (my colleagues at the Unviserity of California Climate Hazards Center and the Famine Early Warning Systems Network, as well as John Magrath, James Firebrace, and Mathilde Berg Utzon). Day in and out, we were helping inform real-world responses to real world catastrophes, yet struggling to keep up as climate change contributed to more extreme hazards. These discussions grew into *Drought, Flood, Fire*. Thank you, friends, for your inspiration.

Next, I would like to thank my lovely wife, Sabina, my adorable children, Amelie and Theo, and my wonderful in-laws, Maurizio and Rochelle Barattucci. This project has required years of long weekends and many late nights locked in the garage. Only the patient support of my family made this work possible – so thank you! I also would like to thank my many coworkers and collaborators, both here in the United States, and in Africa and Europe. I have been incredibly lucky to collaborate with so many brilliant and dedicated people. Our work together has filled my life with purpose and meaning, and sometimes you even laugh at my jokes, which is kind. My friends Joe Peterson, Tana Kincaid, Jim Semick, and Ramzi Hajj helped me believe in this book. Without their support and guidance, *Drought Flood Fire* would have never seen the light of day. I am very grateful for their time, effort, and insights.

Finally, the assistance of Sarah Lambert, my editorial assistant at the Press, provided critical guidance as I navigated the complex process of modern publishing. My remarkable editor at the Press, Matt Lloyd, provided excellent suggestions for modifying the text and structure of this work. I have been very lucky to have had his support and suggestions. I would also like to thank Aleksandr Kats for his thoughtful and thorough copyediting.

1 CLIMATE EXTREMES, CLIMATE ATTRIBUTION, EXTREME EVENT ATTRIBUTION

Introduction

Drought. Fire. Flood. Words that echo through our calamitous past and collective future. Words that threaten to rip our everyday day away, to tear our normal lives apart.

Anticipating, managing, and preventing such disasters has always been a critical test of civilizations. A test of our shared humanity.

Consider, for example, the story of Eregae Lokeno Nakali, shared with me in 2017 by Mathilde Berg Utzon.[1] Mathilde worked as an Africa correspondent for DanChurchAid, a Danish humanitarian nongovernmental organization (NGO) dedicated to supporting the world's poorest people. Driven by compassion, DanChurchAid works to create a more equitable and sustainable world. They provide emergency relief in disaster-stricken areas and long-term development assistance for poverty-stricken communities. Together with Action by Churches Together Alliance, DanChurchAid

---

[1] www.danchurchaid.org/stories/the-global-climate-threat.

works in 130 countries to provide about $1.5 billion in humanitarian assistance each year.

Mathilde and I connected via e-mail in 2017. She was spending a lot of time in East Africa. I spend a lot of time looking at East Africa on my computer screen, working with the US Agency for International Development's Famine Early Warning Systems Network (FEWS NET, www.fews.net). East Africa is one of the most food-insecure places in the world.

In 2017, I had recently started blogging about East African droughts and their relation to climate change.[2]

Mathilde had recently traveled to Turkana County in northwestern Kenya, to document the severe drought that region was experiencing. There she met Eregae Lokeno Nakali (Figure 1.1). Eregage is eighty-one years old. Mathilde writes, "his look is intense and friendly. Between his chin and his lower lip he has a round, silver coloured piece of jewelry. His cheeks are sunken and the wrinkles are deep. It bears witness to a long life. He has experienced how a wealthy life with three wives and enough food has been replaced by extreme poverty and daily uncertainty."

The seminomadic pastoralists of Turkana have evolved a way of life that is, in good times, well suited to their barren surroundings. They raise herds of livestock that can capitalize on sporadic episodes of rain, turning this precious moisture into milk and meat. But Eregae tells Mathilde that in the last year and half there has been almost no rain. He tells Mathilde that he has lost his wife, sister, and daughter due to disease attacking their starved bodies. Eregae and his neighboring villagers have seen their livestock herds destroyed. Eregae takes Mathilde to see the grave of his daughter (Figure 1.2). Mathilde describes the scene:

> Sand coloured, brown and grey stones the size of a fist lie placed in a row in a light brown sandy landscape.

---

[2] Funk C., Concerns about the Kenya/Somalia short rains, October 19, 2016, blog.chc.ucsb.edu/?p=10.

3 / Climate Extremes, Climate Attribution, Extreme Event Attribution

Figure 1.1 Eregae Lokeno Nakali. Photograph by Mathilde Berg Utzon.

Around the stone pattern lie dried branches that the wind has spread randomly across the landscape. This is where Eregae's daughter Napua Eregae lies buried. Only a local person from the area could point out her grave, which is almost invisible in the surroundings.

It was in May last year that her body gave up to hunger. Eregae had no food to give her, and he couldn't afford to take her to the hospital that was too far away. He explains that he couldn't do anything while his daughter died before his eyes.

"It was terrible. I found it very difficult," Eregaye says.

And as though it was not enough for Eregae Lokeno Nakali to lose three of his closest family members, he has also lost his wealth, his identity and employment. All because of the drought and almost no help from outside.

Unfortunately, East African droughts like the ones that afflicted Eregae and Aita are becoming more common. What used to be a one-in-five-years drought now often happens every third year or so. And now warmer air temperatures accompany poor rainy seasons in regions that are already very hot. This heat can exacerbate the atmosphere's ability to pull moisture from the soil and vegetation, desiccating plants and weakening life-sustaining herds of livestock. These more frequent hotter droughts prevent families like Eregae's from accumulating wealth. More frequent disasters sap resilience, enforcing a crushing cycle of poverty. In a poor year these households' herds of goats, sheep, cows, or camels may be destroyed, a huge financial loss for people often living on close to $1 a day.

Eregae and Aita live in the arid Turkana region near the northwestern tip of Kenya. Since Mathilde visited in 2017, the region has been wracked by weather extremes. In the main March–May rainy season of 2018, severe floods in this region and other parts of Kenya displaced more than 300,000 people and led directly to some 132 deaths.[3] Then, in the next October–December 2018 "short" rainy season, the rains came late and were well below normal. Crop production was dismal, about 30 percent of average. Milk production from livestock (a key source of nutrition) dropped by 50 percent. The March–May rains of 2019 were once again below normal, and the region once again faced extreme hunger, conditions right on the border of famine.[4]

As a scientist supporting FEWS NET, these are the poor people I try to protect. Every year, NGOs such as DanChurchAid and government agencies such as FEWS NET and the World Food Programme (WFP) provide billions of dollars in humanitarian assistance to people like Eregae and Aita. These efforts are largely successful at minimizing loss of life. But the number of severely food-insecure people in the world is growing rapidly, and our ability to limit economic devastation remains very limited.

---

[3] www.reliefweb.int/report/kenya/ocha-flash-update-5-floods-kenya-10-may-2018
[4] Kenya Food Security and Nutrition Working Group, "Short Rains Food And Nutrition Security Assessment Findings," Kenya Food Security Meeting, March 8, 2019.

Mathilde and I work at opposite ends of a spectrum of international experts dedicated to preventing, or at least responding to, international disasters in the developing world. Mathilde works with a faith-based NGO helping motivate charitable interventions. I am a hard-core academic at the University of California, Santa Barbara, specializing in using satellites and computers to identify and predict climate hazards. Mathilde works with people. I look at data. We both try to help people like Eregae's young daughter Aita Eregae Nakali (Figure 1.2). I share Mathilde's story with you because it conveys, in ways that my numbers, charts, and maps cannot, the human cost of climate extremes. Climate change is making climate extremes more frequent and intense, not just in Kenya, or in Africa in general, but on every continent. Not in the future, but *right now*.

Every day I work on predicting and monitoring climate hazards. I follow the weather and climate closely, designing and developing information systems that help humanitarian relief agencies save lives and livelihoods. Like a doctor, or a drought detective, I diagnose the factors driving extreme events, so I can improve our capacity to monitor and predict them. I never wanted to study climate change, but it crept into my office like an unwelcome guest, like the coal dust lurking behind a miner's black lungs. Twenty years of data-driven diagnoses brings me to this book. Climate change is hurting people like Aita Eregae.

Over the past few years (2015–2020), the "fingerprints" of climate change have seemed more like a slap. Extreme heat waves, floods, droughts, and wildfires have exacted a terrible toll on developed and developing nations alike. These extremes have impacted hundreds of millions of people and resulted in hundreds of billions of dollars in losses, all across the globe. Fire-afflicted movie stars in California; conflagration ravaged ranchers in Australia; drought-stricken South Africans; poor flooded fisherfolk in Bangladesh; Houston's middle-class families riven by flood – these are just some of the people who have felt the crushing blow of more extreme climate.

But early warning systems can anticipate these disasters, helping to save lives and livelihoods. Such foresight depends on

Figure 1.2 Eregae's young daughter Aita Eregae Nakali. Photograph by Mathilde Berg Utzon

attention and understanding. Viewing the world in an informed way can open the door to a more meaningful and careful life. Perception is a function of both sensory input and our internal conceptions about how the world works. Perceiving the contribution that climate change is making to real-world disasters can clarify our moral, rational, economic, and existential imperatives. My goal here is to provide a data-driven "climate change toolkit" that you can use to see for yourself the influence of climate change on real-world weather and climate extremes. Along the way we will also learn a lot of neat stuff about our beautiful planet. Understanding how our miraculous mothership works can only enrich all our lives.

Exploring Recent Extremes

Since global temperatures jumped up the 2015/16 El Niño, the world has experienced a dramatic increase in the number of extreme events. Climate risks arise through the interaction of climate shocks, exposure, and vulnerability. Climate shocks in turn are comprised of natural and human-induced (anthropogenic) components. So untangling the specific contribution of climate change to multiple extreme events is extremely difficult, and beyond the scope of this book. But we can learn how, in general, climate change is influencing these events. There is little doubt that climate risk is rapidly increasing. Figure 1.3 shows the number of extreme events between

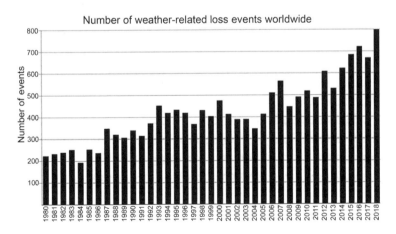

Figure 1.3 The number of extreme events for each year, based on the Munich Re reinsurance company's Natural Catastrophe database.

1980 and 2018, based on the Munich Re reinsurance company's Natural Catastrophe database.[5]

Reinsurance companies insure insurance companies, helping regular insurance companies cope with the huge potential losses associated with severe events like floods and fires. So companies like Munich Re track the number of different weather-related disasters closely. For these companies, such information is critical to their survival. What should be concerning to all of us is that the frequency of such events appears to have more than tripled since the early 1980s. Each vertical bar in Figure 1.3 shows the number of big weather-related catastrophes in each year, beginning in 1980 and ending in 2018. While there are year-to-year variations, we also see a large upward trend. According to this data, the number of extremes has risen from about 200 a year in 1980 to about 800 in 2018.

These catastrophes are not caused by weather alone. A humanitarian disaster – of course - always involves humans, and these weather-related catastrophes arise through the complex interplay of human exposure, vulnerability, natural variability, and human-induced climate change. Still, this time series should really get your attention, especially if you think of any children with affection. These children face an increasingly dangerous adulthood. According to Munich Re, these extremes are accompanied by billions in losses (about \$755 billion [€680 billion]) between 2015 and 2018.[6]

While both news reporters and climate scientists have been very interested in these recent extremes, there has been little in the popular literature addressing these events in an accessible way. That is what we are doing here. This book examines extremes over the 2015–2019 time period, cataloging their severity, and explaining how climate change may have contributed to their intensity or magnitude. The 2015–2019 time period was chosen because this period has been

---

[5] www.munichre.com/en/reinsurance/business/non-life/natcatservice/index.html.
[6] Throughout this book loss estimates will be provided in inflation-adjusted "real" values.

exceptionally "disastrous," and is also recent enough to still resonate with our personal experience. This work focuses on categories of extremes with substantial immediate human impacts (droughts, floods, fires). My goal is to help you both *understand* and *experience* how climate change is contributing to more frequent and intense extreme events.

As we will explore in detail, climate change is already impacting us. Relevant recent examples, which we will explore in this book, include the unprecedented 2017 hurricane season; the 2017 California wildfire season; extreme global temperatures in 2015–2019; the gigantic 2015/16 El Niño event; the sequence of repetitive droughts striking East Africa in 2016, 2017, and 2018; severe flooding in Houston and Bangladesh; coral bleaching in the Great Barrier Reef; and the increase in wildfire frequency and extent in the western United States and Australia. These events have had catastrophic impacts. In 2017, ten Atlantic hurricanes arose in rapid succession: Franklin, Gert, Harvey, Irma, Jose, Katia, Lee, Maria, Nate, and Ophelia.[7] In August 2017, Hurricane Harvey dropped heavier and more widespread rainfall than any other U.S. tropical cyclone on record - more than 60 inches or 152 cm in at least two locations (p. 6 of https://www.nhc.noaa.gov/data/tcr/AL092017_Harvey.pdf). resulting in approximately $85 billion in estimated damages according to Munich Re. In September 2017, Hurricane Irma reached Category 5 on the Saffir–Simpson scale with sustained winds of over 185 miles per hour – for more than 37 hours – longer and stronger than any hurricane on record.[8] Then Maria, another Category 5 storm, devastated Puerto Rico, causing almost 3,000 deaths and leaving millions without power for months on end.

In California, where I live, wildfires ravaged 4.5 million acres between 2015 and 2018, an area almost as big as the state of New Jersey. In 2018, the official cost figures for the Camp and Woolsey fires were $9 billion and $2 billion dollars, respectively,

---

[7] www.nytimes.com/2017/10/11/climate/hurricane-ophelia.html.
[8] www.bbc.com/news/world-us-canada-42251921.

with estimates for all wildfire losses ranging from $15 billion to $19 billion.[9] Estimates for 2017 hover around $18 billion.[10] According to 2017[11] and 2018[12] reports from the risk management/reinsurance company Aon, the global economic cost of natural disasters in 2017 and 2018 totaled $653 billion, the costliest back-to-back years for weather disasters on record. Public and private insurers paid out over $237 billion in 2017 and 2018. In 2019–2020, a staggering and globally unprecedented 21 percent of Australia's forested area burned,[13] as climate change contributed to extreme temperatures and dry conditions.[14] Ecologists estimate that a billion or more animals (mammals, birds, and reptiles) may have perished in these conflagrations.

In 2017, hurricanes Harvey, Maria, and Irma produced some $240 billion in damages. Flooding and Typhoon Hato in China resulted in $15.5 billion in losses. Extreme precipitation and a landslide in Sierra Leone led to the catastrophic death of 1,441 people. Droughts in East Africa pushed some 13 million people into severe food insecurity: these millions of people faced a very real threat of famine.[15] The 2017 drought in southern

[9] www.insurancejournal.com/news/west/2018/11/27/510160.htm.
[10] www.artemis.bm/news/california-wildfire-industry-losses-put-at-13-2bn-by-aon-benfield/.
[11] thoughtleadership.aonbenfield.com/Documents/20180124-ab-if-annual-report-weather-climate-2017.pdf.
[12] thoughtleadership.aonbenfield.com/Documents/20180124-ab-if-annual-report-weather-climate-2018.pdf.
[13] Boer, Matthias M., Víctor Resco de Dios, and Ross A. Bradstock. "Unprecedented burn area of Australian mega forest fires." *Nature Climate Change* (2020): 1–2. www.nature.com/articles/s41558-020-0716-1?proof=trueMay.
[14] van Oldenborgh, G. J., Krikken, F., Lewis, S., Leach, N. J., Lehner, F., Saunders, K. R., van Weele, M., Haustein, K., Li, S., Wallom, D., Sparrow, S., Arrighi, J., Singh, R. P., van Aalst, M. K., Philip, S. Y., Vautard, R., and Otto, F. E. L.: Attribution of the Australian bushfire risk to anthropogenic climate change, *Natural Hazards and Earth System Sciences Discussion*, doi.org/10.5194/nhess-2020-69, in review, 2020.
  www.worldweatherattribution.org/bushfires-in-australia-2019-2020/.
  www.nat-hazards-earth-syst-sci-discuss.net/nhess-2020-69/.
[15] Funk, Chris, et al. "Examining the role of unusually warm Indo-Pacific sea-surface temperatures in recent African droughts." *Quarterly Journal of the Royal Meteorological Society* 144 (2018): 360–383. rmets.onlinelibrary.wiley.com/doi/full/10.1002/qj.3266.

Europe led to some $7 billion in economic losses. In 2018, Hurricanes Michael and Florence battered the United States, while typhoons Jebi and Mangkhut and Tropical Storm Rumbia pummeled China and Japan, with total losses estimated at around $66.4 billion. During July 2018, torrential rains in Japan resulted in almost $10 billion in damages. These natural disasters, combined with the impact of an earthquake, threatened to push Japan into recession. In August 2018, in the Indian state of Kerala, a multibillion-dollar flood led to more than 480 deaths.[16] In Europe, in the summer of 2018, unprecedented wildfires spread as far north as Sweden, and drought-related economic losses were around $9 billion in northern and central Europe. Government and industry are beginning to pay attention.

While humans have always faced the perils of natural disasters, these data suggest that the human and economic cost of climate and weather extremes is increasing quickly as our population and economies expand and our planet rapidly warms. Understanding the link between extremes and warming is both a *moral* and an *existential* imperative. If global warming is increasing the intensity of extremes, then we are all harming people now – drought-affected people like Eregae Lokeno Nakali in Africa, or the nearly 3,000 Puerto Ricans who perished due to Hurricane Maria, the 88 people who died in the Camp Fire in California or the billion animals who perished during Australia's Black Summer.

These increasing perils also pose an existential threat. Already the magnitude of these losses is already sufficient to rattle some of the largest global economies. In poor and moderate-income countries, the toll of the climate crises helps retard economic progress and enforce vicious cycles of poverty. Extreme events can and do contribute to human migration and refugee crises. Understanding the role of climate change can improve our crisis preparation and powers of prediction. If we

---

[16] www.indianexpress.com/article/india/483-dead-in-kerala-floods-and-landslides-losses-more-than-annual-plan-outlay-pinarayi-vijayan-5332306/.

think global warming is increasing the intensity or frequency of a certain type of disaster, then we can better prepare for these events. As we will see later, this preparation often involves a "follow the energy" principle – in other words, we should look for where climate change is creating predictable disturbances to the climate system by essentially 'turning up the volume' of natural variations. Good examples of this are recent improvements to the United States' hurricane forecasting capabilities, or the successful drought forecasts made for East Africa in 2016 and 2017.[17] By learning more about the interaction of climate change and climate extremes, we can face the future better prepared and better informed. One goal of this book is to describe how energy moves through the Earth's energy system, so you can both better appreciate the beauty of our life-sustaining complex planet, and how human-induced warming is altering this system in dangerous and alarming ways.

Extreme Event Attribution and Prediction

This book draws in an informal way from the new science of extreme event attribution. This new science seeks to rigorously identify the "fingerprint" of climate change within weather and climate extremes.[18] Like Sherlock Holmes or an epidemiologist, attribution scientists examine the causes of dangerous events, asking questions like "when Eregae and Aita suffered severe droughts in 2016 and 2017, were these droughts made more intense or more probable due to climate change? Or when the region flooded the following year (in the spring of 2018), was that event made more probable due to climate change?" "Climate attribution" is the accessible process of assigning causal explanations. I work in this research field because it improves my ability to provide effective support for humanitarian relief agencies. I started as a drought detective,

---

[17] www.sciencedaily.com/releases/2018/03/180314144932.htm.
[18] National Academies of Sciences, Engineering, and Medicine. "Attribution of extreme weather events in the context of climate change." National Academies Press, 2016. www.nap.edu/catalog/21852/attribution-of-extreme-weather-events-in-the-context-of-climate-change.

13 / Climate Extremes, Climate Attribution, Extreme Event Attribution

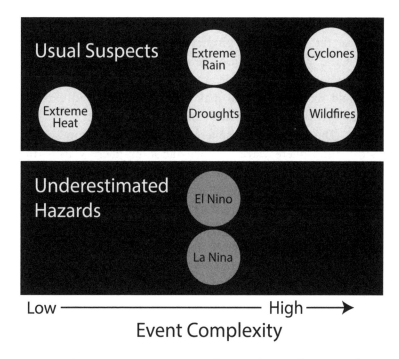

Figure 1.4 Extreme event attribution schema. The top row lists the "usual suspects" discussed in the National Academy of Sciences report on climate extreme attribution (footnote 18). The bottom row lists underestimated hazards examined by the author: El Niño and La Niña, which will be discussed in Chapters 8–10.

trying to understand why rains sometimes fail in Africa. This led to my involvement with climate attribution research. My motivation is humanitarian. If global warming is making an extreme event more common, then we should prepare.

Experts describe (footnote 18) that there is a range in our ability to understand and attribute the role played by climate change in climate extremes (Figure 1.4). At one end of the spectrum, the role that warming plays in heat waves is pretty obvious. As the Earth gets warmer, heat waves get more frequent, warmer, and more extensive in space and time.

Ironically, perhaps, both very dry and very wet conditions (droughts and extreme rainfall events) are the next easiest to understand and attribute. We can be almost certain that global warming is going to make both droughts and extreme rainfall events more common and intense. Why the certainty? Because the same basic physical relationship underlies both tendencies. We can be sure the atmosphere will get warmer. Warmer air can hold more water. The fact that warmer air can hold more water will make droughts drier and extreme rainfall events wetter and more frequent.

Think of the atmosphere as a sponge that grows bigger as the atmosphere warms. When the atmosphere warms, the individual molecules bounce around more, moving farther apart. This produces more room for gaseous water vapor molecules to squeeze in between. This makes it easier for water molecules to move from plant leaves into the atmosphere – a process called transpiration. It also makes it easier for water to evaporate from bare soil. So when conditions are dry and the atmosphere warms, this bigger atmospheric sponge can draw more water from the land. This increases the intensity of droughts. Conversely, when the atmosphere is humid, this bigger sponge will hold more water. Even though we can't see this invisible water vapor, it can be rapidly caught up in storms, lifted, cooled, and sent hurtling down to earth as precipitation. So warmer air leads to more intense rainfall. Another important aspect of these precipitation processes is that atmospheric water vapor moves around with the winds. An individual water molecule may be swept into the atmosphere from a distant ocean, travel thousands of miles, and then converge with H_2O molecules from a totally different part of the ocean, helping fuel extreme rainfall events like hurricanes.

Climate extremes like El Niño and La Niña fall within the intermediate range in our ability to understand and attribute (Figure 1.4). This is where much of my research has been focused, because these "climate events" evolve much more slowly than "weather events" – which means that we can sometimes predict their impacts many months in advance, helping prevent famines.

When El Niño events occur, sea surface temperatures get exceptionally warm in the eastern tropical Pacific. This exceptional persistent warmth can trigger droughts in dozens of countries, some of which (like Ethiopia and Zimbabwe) are very poor and very food insecure. Very recent research, some of it mine, has shown that climate change is making the eastern Pacific sea surface temperature warmer during El Niños. These warmer ocean conditions can trigger more intense droughts in Africa, Asia, Indonesia, and Central/South America. El Niño's sister, La Niña, often follows a strong El Niño event. Ocean conditions in the eastern Pacific cool while temperatures in the western Pacific become very warm. These exceptionally warm west Pacific conditions can trigger severe droughts in places like Kenya and the western United States. Understanding these relationships can lead to important forecasts. For example, in late 2016 we predicted the spring 2017 drought that struck Kenya, Somalia, and southern Ethiopia. This prediction helped motivate early and effective humanitarian responses, helping save the lives and livelihoods of thousands of people. These types of climate extremes begin very simply; we know that the oceans are getting substantially warmer. It then gets more complicated when we analyze how these warming ocean conditions can trigger extreme droughts and precipitation over land.

At the most difficult end of the spectrum, it can be very hard to attribute the role that climate change may play in *individual* wildfire or hurricane events. Humans frequently start wildfires, and once started, wildfires are driven by very complicated interactions between vegetation and highly local weather and winds – which can themselves be driven by wildfires. Hurricanes and cyclones are also rapidly evolving, complicated, nonlinear, chaotic disturbances. Climate scientists have more difficulty modeling these events and their changes. But as we will see, there does appear to be solid observational data linking global warming to more dangerous fires and more devastating hurricanes, cyclones, and typhoons.

Book Structure and Intent

Between 2015 and 2019, climate and weather extremes placed millions of people in harm's way while causing hundreds of billions of dollars in damages. Climate change is making weather extremes more frequent and intense. This book provides an accessible entry point for nonspecialists who want to know how and why. Recognizing *how* climate change is exacerbating expensive and dangerous weather and climate extremes, now, should help us understand why we need to immediately curb our rampant greenhouse gas emissions. But understanding *why* climate change is happening so rapidly may be just as important. Our atmosphere is very thin. Yet our rapid growth in global prosperity is unnecessarily predicated on burning fossil fuels, which threatens the very basis of our fragile life support system. To understand how greenhouse gasses are messing with our planet, it helps to appreciate how awesome the Earth–Sun system is. We absorb solar radiation and turn it into growing complexity. This is our Earth's "fragile flame," a miraculous negentropic order-inducing life support system that we are really messing up.

Addressing the question of *how*, Chapters 2–4 provide a sometimes lighthearted introduction to climate change and climate science. These chapters seek to explain how our Earth–Sun system works, how unique and beautiful it is, and why it is so fragile. We live on a "Goldilocks Planet" where a cascade of energetic balances create excellent conditions for the evolution and sustenance of life. But this life support system depends on a very thin atmosphere and the maintenance of temperatures within a narrow range in which water can take on liquid, frozen, and gaseous states.

Chapters 5–12 then focus on 2015–2019 extreme heat waves (Chapter 5); precipitation extremes (Chapter 6); hurricanes, typhoons, and cyclones (Chapter 7); El Niños and La Niñas (Chapters 8–10); and droughts and wildfires (Chapters 11 and 12). While we can't examine every event, what we find, and summarize in Table 1.1, is that the aggregate impact of these extreme events is massive, from both a humanitarian and an economic perspective. Climate change is already increasing the intensity and frequency of extremes, contributing to dangerous, expensive, and disastrous climate-related crises.

Table 1.1. Notable extremes and impacts examined in this book.

Category	Highlights:
Ch. 5 Temperature Extremes	• Between 2015 and 2019, 59 extreme-temperature disasters, related to 8,800 deaths, 65,592 injuries, affecting 4.4 million people, and resulting in $1.8 billion in losses. • Exceptional warmth, over more than 20% of the Earth's surface, has become the new norm. • These exceptional temperatures threaten the Earth's basic ecosystem services: fisheries, coral reefs, and carbon dioxide-absorbing rainforests. • Without reductions in emissions, global temperature extremes may increase by more than +5°C (9°F). • An analysis of daily global temperature data indicates massive (~15 billion people-days) increases in observed extreme heat exposure events. • The 2050 climate change projections indicate further increases of about 60 to 75 billion people-days.
Ch. 6 Precipitation Extremes	• As air temperatures increase, the atmosphere can hold more water, leading to increases in the intensity of the most extreme precipitation events. • Rainfall observations indicate that global precipitation extremes have already increased by more than 8%. • Between 1998 and 2017, floods, storms, and hurricanes affected more people than any other type of disaster, impacting 2.7 billion people overall and resulting in $1.99 trillion of recorded economic losses.[a] • The 2015–2019 disaster data suggests that the most dangerous non-cyclone storms affected 223 million people, led to more than 9,000 deaths, and resulted in $80 billion in damages.
Ch. 7 Hurricanes, Cyclones, and Typhoons	• According to NOAA data, in 2015–2019, extreme hurricanes, cyclones, and typhoons caused $315 billion in damages in the United States. • Between 2000 and 2019, these extremes caused $746 billion in damages in the U.S. • In 2017, Hurricane Harvey dropped more than 50 inches of rainfall in a few days on the Houston area causing $125 billion dollars in damage.

Table 1.1. cont'd

Category	Highlights:
	• In 2017, Hurricane Maria devastated Puerto Rico. The NOAA-estimated price tag was approximately $90 billion dollars. Deaths: estimated at about 3,000 people.
Ch. 8 Conceptual Models of Climate Change and Prediction	• The amount of energy in the upper ocean is increasing very rapidly. Between 1960 and 1990, the total energy increased by about 3×10^{22} Joules. Between 1990 and 2019, heat content increased by more than five times this amount. • Between 2014 and 2019, the global upper ocean heat energy increase was equivalent to the energy released by about 12 million one-megaton nuclear bombs. • This heat moves around in the oceans, interacting with natural variability to produce potentially catastrophic climate disruptions – and opportunities for prediction. • In October–November 2019, extremely warm western Indian Ocean sea surface temperatures contributed to extreme flooding and locusts in East Africa and drought in southern Africa and Australia.
Ch. 9 Climate Change Made the 2015–2016 El Niño More Extreme	• The 2015–2016 El Niño was associated with extreme drought and air temperatures in Ethiopia, Southern Africa, India, Southeast Asia, Oceania, and Brazil. • These dry arid conditions triggered widespread crop failures, pushing more than 35 million people into food insecurity. • Climate change made the 2015–2016 El Niño about +0.8°C warmer, making terrestrial droughts more intense. • Climate change models predict that more extreme El Niños are likely over the next 20 years.
Ch. 10 Bigger La Niñas and the East African Climate Paradox	• The 2015–2016 El Niño was followed by La Niña conditions, characterized by cool East Pacific sea surface temperatures; but climate change is intensifying the severity of La Niña impacts in some areas by dramatically warming the Western Pacific.

	- In late 2016 and early 2017, East Africa suffered from back-to-back droughts that pushed millions of people into near-famine conditions. - Climate change enhanced the severity of these droughts by increasing Western Pacific sea surface temperatures. - This long-term warming of the Western Pacific may explain the East African Climate Paradox.
Ch. 11 Fire and Drought in the Western US	- In the United States, annual wildfire extent observations exhibit a strong upward trend, with average fire sizes tripling between the early 1980s and late 2010s. - These increases in wildfire extent are tightly coupled with increases in aridity, which are related to both increases in air temperatures and upper-level atmospheric ridging. - The 2017 and 2018 US wildfires have caused more than $40 billion in damages, and more than a hundred fatalities.
Ch. 12 Fire and Drought in Australia	- In late 2019, half of Australia's Kangaroo Island burned, killing more than 17,000 koalas, and more than a third of the island's kangaroos. - In 2019–2020, a staggering and globally unprecedented 21% of Australia's forested area burned. - During this "Black Summer," fires stretched over 186,000 square kilometers (72,000 square miles), destroying over 5,900 buildings (including 2,779 homes) and killing at least 34 people. - Exceptionally warm conditions and climate change enhanced the intensity of the Australian drought and associated fires. - Expert assessments indicate that a billion or more animals (mammals, birds, and reptiles) perished in these conflagrations.

[a] United Nations report, Economic losses, poverty & disasters: 1998-2017, www.unisdr.org/files/61119_credeconomiclosses.pdf.

While we hear about these crises in the news, treating them in aggregate in this book underscores the grave collective nature of this growing peril.

Unfortunately, as we will see in Chapter 13, our current emissions put us on track for calamitous warming. But rising greenhouse gas emissions can also be seen as symptomatic of beneficial growth. Chapter 14 highlights our growing capacity to discover, communicate, and create. Education, technology, and rapid economic growth have lifted billions from poverty. Our ability to image and observe the world has expanded tremendously, and we *can* afford to do the right thing. Between 1961 and 2050, we will carry out humanity's greatest experiment in parallel processing, as billions of individuals grow, think, discover, and consume. We are living in the midst of a potentially *positive* time bomb. The years between 1961 and 2050 will contain as many person-years as 8000–1500 BCE, 1501 BCE–1000 AD, or 1001–1959 AD. Never have so many seen so much, known so much, or done so much – or had such a profound capacity to affect the world for good or ill. We can avoid a global climate catastrophe.[19] But we need to believe in science, believe in each other, and do what is right. This book serves these goals by contributing to a broader understanding of climate extremes in a rapidly warming world.

---

[19] Hoegh-Guldberg, Ove, et al. "Impacts of 1.5 C global warming on natural and human systems." *Global Warming of 1.5° C.: An IPCC Special Report*. IPCC Secretariat, 2018. 175–311. www.ipcc.ch/sr15.

2 WELCOME TO AN AWESOME PLANET
A Series of Delicate Balances Support Earth's Fragile Flame

To appreciate how deeply concerning climate change is, and will be, we need to start by understanding that we are living on a truly amazing life support mechanism that only arose through a series of incredible enabling occurrences. To our knowledge, there is but a single planet supporting life, the Earth. We hang isolated in space, with only an incredibly thin layer of atmosphere standing between us and oblivion. If we could drive our car straight up at highway speeds, we would approach the edge of the atmosphere in a matter of minutes. And into this thin membrane we are dumping about 28 million gigatons of carbon dioxide every day.[1] The rational decisions of nearly 8 billion people are resulting in collective insanity as we choose to destroy the delicate balances that support Earth's fragile flame.

Scientists tell us that life on Earth has only been possible due to a series of delicate balances that arise at all scales of the universe (Figure 2.1). Physicists often refer to this as the "anthropic principle." There is a lot debate on this topic. Some scientists believe that the structure of the universe is inherently conducive

---

[1] www.globalcarbonproject.org/index.htm.

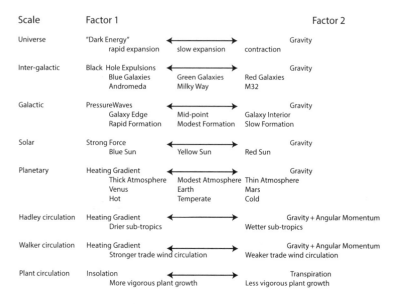

Figure 2.1 Reality as a series of fortunate compromises. At all scales, the evolving structure of our universe is always composed of a set of balancing forces.

to the evolution of complexity, life, and intelligence. Others believe the "miraculous" life-supporting structure of the universe is simply due to "selection bias," i.e., a universe has to be pretty similar to this one to support life in order to be observed. While this issue is keenly debated, and extremely hard to prove one way or the other (since we only have one universe as our scientific "sample"), there is no debating the fabulous series of fortunate balances that have supported the development of intelligent life on Earth. A brief review of these balances will provide context for our later discussions of climate change.

Enabling Occurrence 1: Balanced Dark Energy and Gravity

Throughout this book we will encounter complex dynamic systems. While tremendously varied, these systems

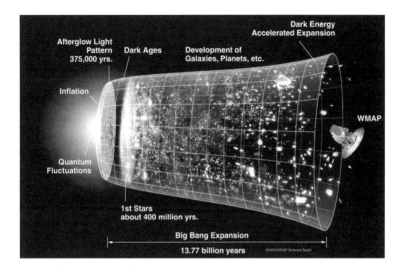

Figure 2.2 A figure from NASA depicting the temporal evolution of the universe, as seen by the Wilkinson Microwave Anisotropy Probe (WMAP). At the far left, the Big Bang. To the far right now. map.gsfc.nasa.gov/media/060915/index.html

always involve multiple balancing forces that evolve across space and time (Figure 2.1). At universal scales, dark energy and gravity play a cosmic tug-of-war. The still poorly understood "dark energy" drives the universe toward rapid expansion. In contrast, gravity pulls on the fabric of space-time, tugging toward contraction. At different stages of the universe, dark energy seems to have had the upper hand, while at others, gravity ruled. Informed by satellites, computer models, and decades of research, astronomers have assembled an exciting view of the temporal evolution of the universe (Figure 2.2).

Some 13.7 billion years ago, at the time of the Big Bang, some "quantum fluctuations" are thought to have occurred in a universe that may have started out as a single dimensionless point. Astrophysicists posit an incredibly small dense hot place under incredible pressure. Then the quanta got queasy, and blammo, in less than one-quintillionth of a second (1^{-32}) the universe underwent a cosmic inflation. Ripples in space-time formed the seeds of latent intergalactic structures. At just one-millionth of a second we began to see the formation of protons and neutrons. Mind you, average temperatures were still about 1 billion degrees Celsius. This Big Bang caused the size of the universe to increase incredibly, leaving behind the background radiation afterglow that satellites see today in all directions of the cosmos.

From school you might remember the universal law of gravitation: $F = M(m_1 + m_2)/d^2$, where M is the universal gravity constant, m_1 and m_2 are the masses of attractive bodies, and d is the interbody distance. What you may not know is that a small variation in the magnitude of M would radically alter the history and shape of our universe. Increase M by 10%, and the Big Bang that birthed our universe may have already concluded with a big crunch of coalescing and colliding stars. Decrease M by 10%, and energy and matter may have become so widely dispersed that no coherent structures could have formed – no stars, galaxies, light, or life.

But without the repellent force of dark energy, the Big Bang would have never happened. According to the current "standard model of Big Bang cosmology," the force of gravity is offset by "dark energy" – an elusive, extremely sparse type of vacuum energy pervading every place in space. While great uncertainty surrounds this new area of cosmology, most astronomers now believe that something (dark energy) exists everywhere in space, and is everywhere exerting a very weak negative pressure on the space-time manifold. This negative pressure counteracts gravity, pushing the universe apart. So at the most massive cosmic scale, scientists have found a delicate balance between dark energy, with a tendency toward expansion, and gravity, which produces contraction and compression.

Just as space-based infrared satellite observations of distant stars in the 1990s helped fuel a revolution in cosmology, leading to an ever deeper view of cosmological time, so twenty-first century space-based microwave observations helped define and explore the evolving shape of our Universe. By measuring the radiation emissions associated with the Big Bang, the newest

NASA satellites are tracking the expansion of the universe. Everything with a temperature emits radiation with a characteristic frequency. Super-warm things emit very high-frequency radiation (gamma rays and x-rays). Medium warm things like our sun emit mostly medium-frequency radiation – visible light. Very cool things emit radio and microwaves.

The universe is very cool, but not as cool as it could be. Observed microwave radiation emissions from distant empty locations are about $-269.43\,°C$ (2.725 K), close to absolute zero, but *not* absolute zero. When satellites peered around the universe, looking for this faint background radiation, the spatial distribution of these background emissions temperatures was *isothermal* – pretty much the same in all directions. This isothermal background has been interpreted as the leftover heat from the Big Bang.

As matter coalesced, the relative impact of dark energy decreased, and the rate of the universe's expansion slowed to a near constant rate. It is estimated that only very small changes in either the strength of the universal gravitational constant (M) or the negative pressure force associated with dark energy would have led to the early destruction of our universe. A little dash more M, and the universe would crush in upon itself. A little more dark energy, and all celestial objects would end up hurtling away from each other at speeds beyond the speed of light, cut off forever from even communication and observation. Each object would be ultimately alone in a universe that appeared utterly black.

This, thankfully, was not to be. After about 400 million years, some pockets of matter became dense enough to support nuclear fusion. Stars began to form, and our longest Dark Age came to a close.

Enabling Occurrence 2: A Green Galaxy

As we go from intergalactic to galactic, to solar, to planetary, and then to biotic scales, life (defined by our sample of one-planet Earth) appears to spring up where a delicate balance between forces exists (Figure 2.1). At the grandest scales, gravity contracts, while dark energy expands, our universe. Then within and between galaxies, gravity is offset by black hole expulsions and the pressure waves of rotating matter. Then within a star, the competition between gravitational attraction and the quantum-scale strong force determines solar density and magnitude. The strong force refers to the repulsion between protons at very small distances. Then with our planet's oceans and atmosphere, vertical and horizontal temperature gradients balance insolation, producing our terrestrial climate. Within plants, the offsetting effects of solar insolation and transpiration produce the energy supporting all terrestrial creatures.

Just as contrasts of dark and light create art, or contrasts in pitch and tone create music, most physical systems need contrasting forces in some dynamic quasi-balance. Scientists find these balances beautiful, and our inquiries into them fueled the seventeenth century Scientific Revolution. In 1609, Kepler plotted the orderly sweep of the planetary spheres. In 1638, Galileo measured the attraction of gravity. In 1668, Isaac Newton connected these scales in one of the most breathtaking leaps of intellectual thought. He realized that the same gravitational force that describes the motions of objects on the Earth also controls the planets in our solar system, resulting in one universal law of gravitation. The same force drives the dance of planets and my own body's descent as I watch the bicycle handlebars pass below me as I tumble past. But as we shall see, gravity is a creative force. Understanding why can make our lives more meaningful, careful and wonderful. The universe is neither cold nor barren, nor is she likely to coddle and forgive. While the Earth seems special to us (and it does seem fantastically unique and complex), it is also just another chunk of rock in space. Like nebulae and galaxies, we all have to follow the same rules, the laws of physics. The arc of the cosmos from the Big Bang to our final end resembles in some ways the arc of a human life. So often what we perceive as stasis is in fact the result of a delicate balance of forces that can forge complex and evolving patterns. But so far the push and pull of galactic gravity has been "just right."

At intergalactic scales one sees this interplay in blue, green, and red galaxies. Astronomers believe that many galaxies have black holes at the center. In blue galaxies, like Andromeda, a massive black hole at the center pulls in the surrounding matter, creating rapid rotation speeds while emitting large quantities of high-energy gamma radiation. At the other end of the spectrum, the center of "red" galaxies such as M32 are thought to contain small black holes. They rotate slowly, and exhibit less radiation extremes. Within the middle of the galactic scale, the Milky Way seems to occupy a "just so" happy medium, with a black hole at the center big enough to keep things mixing but not large enough to send gene-smacking gamma radiation all over the place.

Enabling Occurrence 3: Good Galactic Real Estate

Within galaxies we see an opposition between gravity and pressure waves. Pressure waves are generated by galaxies' rotation and the energy emissions of stars. Near the edge of galaxies, large rotational speeds tend to support intense pressure waves and rapid levels of star generation. Near the center of galaxies, rotational speeds are slow, and weak pressure wave activity leads to slow levels of star generation. Near the center of galaxies, where we find our sun, conditions appear to be commendable for life formation. Gravity and pressure wave activity is strong enough to support star formation, but the pressure wave intensity is not so severe as to create frequent collisions and disruptions.

Within our galaxy we see the opposition of gravity (pulling in) and effects of rotation and radiation (pushing out). Once again, it seems that Sol's position is close to "just so": just close enough to the Milky Way's rim to get a decent dose of high-molecular-weight elements (like carbon, phosphorous, and oxygen, which are formed by fusion in large stars) but just far enough from the center that tidal gravitation variations haven't pulled us apart.

Enabling Occurrence 4: A Very Nice Star

Within stars we see the opposition of gravity (pulling in) and effects of the strong force (pushing out). The strong force is the repulsive force released when neutrons and protons are forced together during the process of fusion. Some new or massive stars have very high rates of fusion, and appear to us as blue in telescopes. These stars typically have a much greater solar mass, sustain fusion at much higher temperatures, and release much more gamma radiation. Some old or small stars, on the other hand, appear to us as red and burn at lower temperatures. The temperature of a star may have trade-offs when it comes to supporting life. Cool red stars release less gamma radiation, which can cause mutations, but they also emit less potentially life-supporting visible light. Blue stars may occupy the other extreme. Yellow stars, like ours, may provide a happy medium, at least for life similar to Earth's.

Enabling Occurrence 5: A Very Nice Atmosphere

And our Earth itself seems pretty special. At its core, molten nickel revolves, creating a giant magnet, which in turn creates a belt of radio waves (the Van Allen belts), which keeps our precious genetic heritage safe from descending gamma radiation death rays. Without the Van Allen belts, genetic mutations would likely have been so common as to make life impossible. The Earth also has a fine assemblage of elements, a reasonably stable atmosphere and climate, and lots and lots of water, in all three forms (liquid, gas, and frozen). As far as we know, such planets are very rare.

Our atmosphere may also be very conducive to life. Way up high in the stratosphere, a layer of ozone helps protect us from ultraviolet radiation. At the same time, the nitrogen and oxygen that comprise the bulk of our atmosphere are transparent to the solar radiation beaming down on us from the Sun. Evolving life loves to get all that high-quality incoming energy. Our atmosphere, ocean, and distribution of land also play an important role in making things interesting by producing complicated patterns of climate. These complicated climate crenulations support numerous niches in which diverse life forms can develop to consume the Sun's bounty.

The Sun's steady input of energy allows for complex systems, motions, and emotions to evolve on the Earth. We get weather and ocean currents because we are differentially heating the bottom of the atmosphere and the top of the oceans. There are also important north-south differences in heating. The equator and summer hemispheres receive more energy, and this gradient of heating drives atmospheric and oceanic circulations.

It turns out that the amount of atmosphere a planet maintains has quite a bit to do with a planet's mass and distance from its sun. Too small and too close, and the combination of weak gravity and strong solar wind can cause the atmosphere to slowly blow away. This has the effect of cooling the planet, because the atmosphere tends to absorb and re-emit energy. This process of absorption and re-emission, called "radiative transfer" by atmospheric scientists, is also called the "greenhouse effect." While there have been some crazy variations in carbon dioxide levels over the Earth's nearly 4.5-billion-year history, they have been pretty stable for about the last 800,000 years (Figure 2.3), ranging for the most part between approximately 170 and 280 parts per billion. These carbon dioxide levels, together with the complex influence of clouds and the redistribution of energy by our oceans and atmosphere, have kept temperatures in a magical range that embraces a miraculous substance – water – in all its phases: liquid, gas, and solid. Gaseous water transports moisture from the oceans onto land, sustaining terrestrial life. Liquid water teams with aquatic flora and fauna. Ice caps help reflect some of the sun's rays, keeping us reasonably cool.

The Earth occupies a fortuitous middle ground in the atmospheric arena. We have enough atmosphere of the right kind to protect us from the worst impacts of high-energy radiation and enough of a greenhouse gas effect to bring us up to a comfortable average temperature. Temperatures conducive to

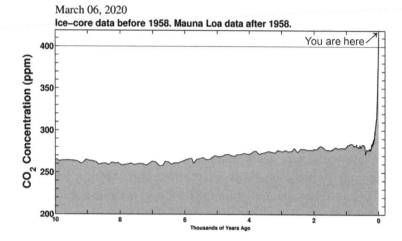

Figure 2.3 Atmospheric carbon dioxide levels from ice cores and the Mauna Loa Observatory. Data between 800,000 years ago and 1957 are based on ice cores. Values from 1958 to the present are based on observations at the Mauna Loa observatory. Image courtesy of Scripps, accessed September 28, 2019. scripps.ucsd.edu/programs/keelingcurve/wp-content/plugins/sio-bluemoon/graphs/co2_800k.png

rainfall, rivers and reflective ice caps. This can be contrasted with planets like frigid Mars, with almost no atmosphere, and Venus, where we find an intense greenhouse gas effect and extremely high temperatures.

The current manmade increase in CO_2 is also shown in Figure 2.3. This figure combines very long estimates of carbon dioxide levels derived from polar ice core estimates with the 1958–present "Keeling Curve," based on continuous measurements taken at the Mauna Loa observatory in Hawaii. The Keeling Curve is named after Charles David Keeling, who began monitoring CO_2 at Mauna Loa in Hawaii in the late 1950s. The current jump beyond 400 parts per million is way beyond the range of natural variations observed over the last 800,000 years. We are all on a rock-and-roll ride of unprecedented proportions.

Pause for a moment to ponder Figure 2.3. Before the end of this book you should understand very well that we are on a speck of rock in a void in space with a wafer-thin atmosphere, chain-smoking unfiltered cigarettes and singing along to the radio as we slam down the gas pedal hurtling our bus into Dead Man's Curve.

Enabling Occurrence 6: The Hadley and Walker Circulations

If all the energy arriving in the tropics stayed there, the Earth would be much more boring, and most places would be a lot cooler. Luckily, we have a set of interconnected global circulation patterns, the Hadley and Walker Circulations, which redistribute heat from the tropics. Climatologists refer to the north–south average climate conditions as the Hadley Circulation. The origin of the Hadley Circulation is the differential heating of the tropics. This creates a band of low pressure near the equator (Figure 2.4). Winds are drawn toward this low pressure. As they move towards the equator, however, the Earth rotates beneath them, from west to east, causing the apparent westward motion of the trade winds. Near the equator the northern and southern trade winds converge, creating vertically ascending motions, often accompanied by rainfall. In the upper troposphere (more than 10 km above the Earth's surface), these winds move towards the poles and curve east, forming the jet streams, and feeding into bands of upper-level convergence near 30° south in the southern hemisphere and 30°

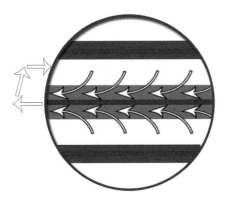

Figure 2.4 The Hadley Circulation.

32 / Drought, Flood, Fire

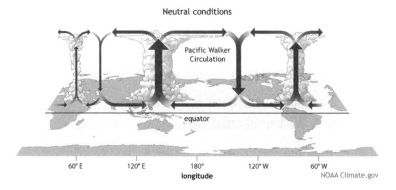

Figure 2.5 Generalized Walker Circulation (December–February) during ENSO-neutral conditions. Rainfall associated with rising branches of the Walker Circulation is found over the area surrounding Indonesia, northern South America, and central Africa. Credit: NOAA Climate.gov drawing by Fiona Martin. Downloaded from Tom Di Liberto's blog, www.climate.gov/news-features/blogs/enso/walker-circulation-ensos-atmospheric-buddy, on September 28, 2019

north in the northern hemisphere; where at the surface we often find subtropical high pressure cells and arid regions like in the western United States.

We can build up a more detailed picture of how the global average circulation works by including east–west circulations along the equator, typically referred to as the Walker Circulation (Figure 2.5). The Walker Circulation was named after one of my personal heroes, Sir Gilbert Walker, the British Royal Meteorologist in India during the early 1900s. Motivated by the terrible Indian droughts and famines at the end of the nineteenth century, his research into Indo-Pacific climate variability led to his discovery of the "Southern Oscillation," the atmospheric component of the El Niño–Southern Oscillation (ENSO). In 1896–1897 and 1899–1900, India experienced severe droughts that produced some of the most severe famines on record. In 1896–1897, some 5 *million* people are thought to have died in British-controlled districts, and 1 million are estimated to have perished in 1899–1900. In 1904, Walker, a British mathematician familiar with fancy new empirical analysis techniques such as correlation, entered the British Colonial Service as Director of the Indian Meteorological Observatory, with the goal of predicting Indian Monsoon variations.

Combing through the worldwide weather records, Sir Gilbert discovered the Southern Oscillation. The Southern Oscillation captured the "covariability" of the sea level pressure variations near Darwin Australia and Tahiti. Under normal conditions (Figure 2.5) higher pressure over Tahiti and lower pressure over Australia helped drive an east–west wind pattern associated with heavy rainfall and ascending air in the area surrounding Indonesia. This pressure differential helped drive the east-to-west surface winds that sweep across the equatorial Pacific. These winds feed heat and moisture into the area around Indonesia, fueling intense precipitation and rising air.

During an average year, this region (10°S–20°N, 70°E–180°E) receives a whopping 82 inches or 2 meters of precipitation. This much rain across 4 million square kilometers makes a cube of water 43 km on each side. If all this water were to condense at once, it would release about 184 billion terrajoules of energy – roughly equivalent to 184 billion big nuclear bombs going off at once. We receive a tremendous amount of energy from the Sun, and quite a bit of it ends up evaporating water from the oceans, and then much of this energy ends up being released as precipitation over the area surrounding Indonesia, where the eastern Indian and western Pacific oceans meet. This region of exceptionally warm ocean waters is referred to as the Warm Pool. This energy rises up, is carried by eastward-flowing "westerly" winds in the upper troposphere, and is then lost as radiation to space as the upper-level winds travel east. This cooling, combined with relatively cool sea surface conditions along the eastern coasts of the Americas, causes subsiding motions that help produce the dry subtropical high-pressure cells to the west of Mexico and Peru. These high-pressure cells, in turn, help drive the westward-flowing "easterly" trade winds that feed moisture into the Indo-Pacific Warm Pool. Together, the components of the Walker Circulation comprise a giant "heat engine" that keeps the general circulation of the earth circulating, with all of the energy, ultimately, coming from the Sun.

As we have seen with astronomical phenomena, the strength of the Hadley and Walker Circulations arise, again, through a balance of forces. The Hadley Circulation is largely driven by the strength of the gradient in surface warming between the poles and the equator, along with the speed of the Earth's rotation. This rotation feeds the "apparent" Coriolis force, which turns equatorial-headed winds to the west (in the Northern hemisphere, as in Figure 2.4). Given a stronger heat gradient, or stronger rotation effects, we would expect to see stronger trade winds near the equator, and stronger westerly winds in the mid-latitudes. The influence of heating and spinning are not identical. Spinning, in general, tends to produce "banding"; when we look at images of Jupiter or Saturn, for example, the bands of these large planets are formed of counter-rotating longitudinal cells (bigger planets tend to have higher surface rotation speeds).

If the Earth's rotation slowed sufficiently, we would expect to have just one boring vertically overturning circulation. At all longitudes, air would meander down toward the equator, warm, and rise. Then, near the top of the tropopause, it would move to the poles, sinking somewhere like Manitoba or the Central Siberian Plateau. Such a unicellular circulation would produce much greater north–south climate extremes. At present, mid-latitude winds and ocean currents transport a great deal of heat to Western Europe and America. This mixing of warm and cool waters keeps much more of our planet in a nice temperate regime. Conversely, increasing our rotation speed would create more counterrotating bands, like we see on Jupiter or Saturn, potentially disrupting ecosystems, habitats, and migration patterns. Fortunately, weather patterns on Earth have been, for the most part, conducive to the rapid development of life.

Enabling Occurrence 7: Amazing Life

To the long list of "just so" life-supporting features associated with our world, we might also add expeditious location and timing (Figure 2.6). We find ourselves in a nice part of

Figure 2.6 Universal timeline showing the background radiation released by the Big Bang, a nebulae, a galaxy, the Earth, and stromalites in Australia.

the Milky Way Galaxy: not too close to the center where we are likely to get fried by high-energy radiation being emitted by the galactic core, but not so far out on the fringes that the density of energy and matter is terribly sparse. The Earth formed as a chunk of rock some 4.5 billion years ago, and then slowly cooled. The first evidence of life (that we know of) appeared about 3.8 billion years ago, in the form of small strands of organic carbon now embedded in sedimentary rocks. These compounds may indicate early life forms living near thermal vents. They could have used the vents as a source of energy. There does seem to be a good case to be made for a very early evolution of life, perhaps very early indeed in Earth's history, but since the oldest rocks on Earth have since been melted, the evidence has been erased.

Over time, the atmosphere of the planet has changed a lot. At first, the atmosphere had a high concentration of carbon dioxide because of all the intense volcanic activity during the Earth's first billion or so years. Then CO_2 levels dropped, and oxygen increased, because of the rise of plant life and photosynthesis. Atmospheric CO_2 was also locked up in sedimentary rocks, like limestone and coal, which are themselves produced by life. In other words, life on Earth dramatically changed the atmosphere on Earth, making the Earth cooler and more hospitable, at least to animals like ourselves.

Each day, plants perform a subtle dance with desiccation and photosynthesis. If we zoomed in to look at a leaf's surface, we would find tiny holes, called stomata, through which the plants exchange water vapor and CO_2 with the atmosphere. Generally, plants get water from the ground and CO_2 from the air, and then use the magic of photosynthesis to make carbohydrates and sugars. The problem plants face is that stomata are either open or closed, and water and CO_2 flow through the same channels. So when they open the doors to draw in CO_2 to fuel photosynthesis, the plant's precious vital fluids (water) are also escaping through the same door. So again, we have a balance. Plants definitely want enough nice yummy heat energy (incoming solar radiation) to drive photosynthesis, but (especially in semi-arid regions) they want conditions cool enough to limit water loss. Our current temperature conditions provide wide regions of the Earth where that balance can be satisfied.

Welcome to an Awesome Planet

What does the very early evolution of life on Earth mean? It is very hard to be certain, given the small size of our sample (1), but maybe it means that life, as an organizing principle that produces self-replicating, evolving, intelligence-auto-augmenting organisms with a genetic attraction to microwave popcorn, may not be as rare as we might think, given the "right" conditions.

In the European fairy tale of Goldilocks,[2] a young girl wanders into the forest and stumbles into a house owned by bears who are out for the afternoon. She examines the chairs/breakfast/beds, always finding a middle version that is "just right." As our understanding of our Earth and universe has evolved over the past fifty years, what seems remarkable to many scientists is how "just right" the universe and Earth seem to be. At many scales, the details of the universe seem

---

[2] Credit for this discussion goes to Jan Zalasiewicz and Mark Williams in their excellent book, *The Goldilocks Planet: The 4 Billion Year Story of Earth's Climate* (Oxford University Press). Fred Adam's *Origins of Existence: How Life Emerged in the Universe* (The Free Press) also informed this discussion, as did several popular articles by Caleb Scharf on the important role black holes play in maintaining galactic structure.

surprisingly conducive to life, with the complex structures supporting life arising through a delicate interplay of offsetting forces. In a myriad of ways we seem to live in/on a universe, galaxy, solar system, and planet that are extremely well suited for habitation. One so inclined could see this as evidence for a divine plan or creator. Another interpretation has to do with "selection bias." While other configurations for our universe/galaxy/solar system/planet are very likely, such configurations are unlikely to support life, and hence be observed. So the uniquely life-supporting qualities we observe may just be due to the fact that we are around to observe them. Yet if you wanted to believe that a benign Creator, or a divine evolutionary process, acted to produce a universe conducive to the creation of life, that would also seem to fit the facts just fine. Seventeenth-century scientists, such as Galileo, believed that the Earth was like another Bible, where God had written his beautiful and mysterious messages for us to discover. I don't think exposure to the past four hundred years of tremendous, astounding, and beautiful scientific discoveries would cause Galileo to lose faith, but rather to double down, stunned by wonder.

We only have one planet. Life appears to have formed on this planet just about as quick as possible, and then exploded in awesome and amazing ways to fill a crazy number of niches with fantastic and beautiful creatures. Scientists estimate the number of species currently populating Earth at around 8 million, with perhaps about 7 million more still waiting to be discovered.[3]

So welcome to an awesome planet. Going back over our checklist (Figure 2.1), we find that a nice balance of gravity's pull and dark energy's push resulted in a rich quilt of galaxies and stars, with us ending up in a comfortable green galaxy, in a prime mid-spiral section of the Milky Way, rotating around a nice sun (not too close, not too far) gently emitting in the yellow

---

[3] Most of these are simple eukaryotic creatures; www.nature.com/news/2011/110823/full/news.2011.498.html.

part of the energy spectrum. A healthy quantity of atmosphere and greenhouse gasses, along with the Van Allen belts, keep out most high-energy particles and maintain a reasonable temperature. The Hadley and Walker Circulations pitch in, creating clockwise-rotating circulation cells in and over the Pacific and Atlantic Oceans, transporting heat away from the equator and depositing it in poleward latitudes. These circulations bring life-giving moisture to the continents, supporting abundant life.

Over time – and this is very important – life has acted in such a way as to sequester (reduce) the amount of atmospheric carbon dioxide, which was very high in the hot heady days of Terra's volcanic adolescence. Plants and marine organisms extracted tons and tons of carbon dioxide through photosynthesis and the creation of shells and skeletons. These extractions had a major influence on our atmosphere; an effect that is being undone at a rapid pace as we burn and release tons and tons of sequestered coal, gas, and methane into our atmosphere (Figure 2.2).

Pulling back, what we see at all levels of our world's structure, written backward and forward across the face of time and space in a thousand different ways, is the complex interplay of forces trading matter and energy in delicate systems that can, sometimes, evolve and innovate. During the sixteenth and seventeenth centuries, visionary scientific revolutionaries read the word of God written in the bones of the world. Today, this grand adventure bears noble fruit, uncovering a rich tapestry of complex inventive processes at every scale.

Note, however, that the manual for Spaceship Earth does not contain a warranty. Right there on the cover, next to the red "Don't Panic" logo, is emblazoned "No Returns If Opened." The universe provides no guarantee that we can't seriously fudge up this miraculous lifeboat.

For religious communities that trace their roots back to Genesis, I would suggest that humanity's original sin and concomitant free will provide little assurance of divine intervention. God promised to never send another flood to destroy the world, but I doubt She will intervene if our greed, neglect, and willful ignorance leads to Her nice planet's destruction.

Life has already dramatically changed carbon dioxide levels and the level of greenhouse gas warming during out planet's history. And we are certainly doing so again at a tremendous rate (Figure 2.3). If a divine entity really did go to all the trouble of helping produce a benign universe for human life, we are trying very hard to reach out of the cradle and stick our finger in the power socket. But a key concept here is balance amid a delicate array of fluxes and forces. This is a challenging, humbling, but beautiful insight provided by science. One of our first scientists, Heraclitus, wrote the words '*Panta Rhei*' (everything flows) in about 500 BC. The more we learn about nature, the more miraculous it is that everything doesn't just fall apart. I hold out my hand. Flex my fingers. Ponder the immediate complexity of the neurons firing, the coordinated movements of the muscles and tendons. Think about the release of energy by adenosine tri-phosphate (ATP) that fuels my muscles. Each day, my cellular mitochondria, which probably exist due to a eukaryotic host cell's happy capture of bacterium in my far distant past, produce the equivalent of my body weight in ATP, the principal energy currency of my cells.[4] At longer time scales, a molecule of water in my body might reside for 12 days. Most of the cells in my body, except for my brain, die and are replaced. Each of us is more "fluxed" than "fixed," and it is wonderful that our bodies can often work so well. I am very glad to be a small part of this amazing planet. Like a standing wave in a river, water, energy, matter, and life flows through me, through us all, binding us together on the bright side of empty.

---

[4] Törnroth-Horsefield, Susanna, and Richard Neutze. "Opening and closing the metabolite gate." *Proceedings of the National Academy of Sciences* 105.50 (2008): 19565–19566.www.pnas.org/content/105/50/19565.long.

3 THE EARTH IS A NEGENTROPIC SYSTEM, OR "THE BRIGHT SIDE OF EMPTY"

It's funny, but of all the profound thoughts, sights, experiences, and insights from my fifty-four years, I seem to remember most clearly those times I almost died through my own acts of stupidity. Damn self-serving, embarrassing, self-preserving biology. When *I* design a replacement species, it will remember my good jokes, which unfortunately arise less frequently than episodes of attempted self-de-existification.

One of the most notable of such episodes arose decades ago, while I was still in serious bachelor mode. Together with a bunch of close friends I went canoe camping in Michigan. There was afternoon beer drinking involved, and afternoon whiskey drinking as well. And occasional canoeing. So when we stopped to make camp, I was not at my "woodsman" best as I went to collect firewood. Not when I broke the fallen rotten log apart. Not when the nest of wasps winged forth seeking vengeance. Not when my alcohol-addled brain failed to instruct me to run. Only when my spinal cord took over control did the faintest signs of intelligent life emerge, pumping my legs to run my body through the quickly darkening woods as many wasp stings came in quick succession.

The Earth Is a Negentropic System, or "the Bright Side of Empty"

Back at the campfire everything was just fine. The tinder caught, growing flames flickered. Yet venom coursed within my veins, and suddenly without warning I bent at the waist, pitched forward, and planted my face in the woodland duff. Anaphylactic shock then came quickly, my lymph nodes swelling, cutting off the blood flow to my brain. John and Ramzi went for help. Jack and Steve waited. I gave Steve my pocket knife and told him he might need to give me a tracheotomy. He was really not into that idea. More swelling. I figured I was pretty much done for. I felt pretty okay with that. My mind was at peace, ready to move on, but some beautiful parting thoughts blossomed in my mind.

I could see history stretching around us – backward and forward in time – like a black rubber mat – flexible, resilient, and yet markable – memorable. Where we placed our feet mattered, creating channels of propensity and ordered progression toward a (hopefully) better tomorrow. Like synaptic pathways in our brains, our actions left a residue across the manifolds of space-time, making it more likely that others followed in our footsteps, wearing in paths toward a future good or ill depending on the cumulative effect of our individual decisions. I thought this would be a nice truth to describe in soliloquy as I exited stage left. Feverish, faint, flushed, and feeling drawn down a gentle river that flowed in a pink marble cave toward a glowing light growing closer and closer, I mouthed my final words of wisdom – which came out as an eloquent "mrrrmmppphh," since my tongue and glands were so engorged. I tried to convey how our individual steps matter, having mass, helping mold the path of future human time... but all that came forth were guttural groans. Ignominious indeed. Poor Steve had to sit there with the knife trying to decide whether or not to – quite literally – cut my throat.

I was saved by a Rainbow Gathering just around the river's bend. Or more precisely by some local high school heroes at the Rainbow Gathering who rushed me to the hospital and probably saved my life. Rainbow Gatherings are "temporary loosely knit communities of people who congregate annually in remote forests around the world for one or more weeks at a time to enact a supposedly shared ideology of 'peace, harmony, freedom, and respect.'"[1] My friend John rushed to the Rainbow Gathering, desperately seeking assistance. Most of the attendees were no help – too inebriated or high to provide assistance. But some teenagers from the town next door generously came to my assistance. They put me in the back of their car and plowed through the woods, getting me to the emergency room just in time. I was sliding down that pink marble tunnel closer and closer and closer to the light.

This chapter flows from that incident, that failed opportunity to tell a profound truth, the new lease on life, the ensuing opportunity to go to graduate school and study how this beautiful planet works. Given a few hours left to my life, a candle guttering and sputtering, a pencil and paper, this I would write.

Principle 1: Emptiness $\neq$ Bad

Most of us, deep down, or even not so deep down, feel a profound sense of panic when confronting our astronomical, molecular, and psychological isolation. Behind so many of our frantic efforts to produce, consume, attract notice, and fundamentally matter, lies fear of the abyss, the void, and the end of our existence. We look at the stars at night. We ponder the big empty. We think about the incredibly insane distances between the stars. We feel very alone, very small, and very insignificant. We do the math, deducing perhaps nihilistic despair. Take the sum of all the stuff (stars, planets, quasars, pulsars, gas, clouds, nebulae, etc.) and divide it by universal volume, and we end up with a number very very close to zero [$\varepsilon/\sim\infty$ = bupkis]. Such is the logic of insignificance. If science tells us that we are not significant, then perhaps science itself is suspect.

But maybe dividing by the volume of the universe or a galaxy is dumb math. Who cares about a bunch of nothing?

[1] en.wikipedia.org/wiki/Rainbow_Gathering.

43 / The Earth Is a Negentropic System, or "the Bright Side of Empty"

Naïve geometry can invoke stupidity, injecting human scales where they don't belong. Here, on this planet, we have the feel of wind, the scent of air; the pigeon/wren/thrush/(insert bird name here) taking flight in the morning. Not (ever) to forget the smell and taste of coffee. This can be written in the language of science.

$$\frac{\textit{Thrush taking flight in early morning}}{\textit{Volume of nearly infinite universe}} = \textit{Thrush taking flight in early morning}$$

What if lots of nothing was needed to produce singular life, a quietly sleeping child? If you can share with me a few moments pondering the creative necessity of the void, the Big Empty that surrounds us, you might look out upon the night sky with greater wonder, and rest assured that the universe is more benign than you believed.

The Big Empty helps overcome entropy and the universal tendency towards heat death. Entropy is a measure of the universe's organized thermal structure and ability to do work. The early discovery of entropy and its implications has been associated with insanity and despair. The idea was first advanced by Nicolas Léonard Sadi Carnot (1796–1832) in his *Reflections on the Motive Power of Fire* (1890). Carnot introduced the idea that the efficiency of a steam engine will be proportional to the difference in temperatures between the hot and cold reservoirs, an idea that was later formalized into an expression for the thermodynamic efficiency of a system:

$$efficiency = \frac{(T_{hot} - T_{cold})}{T_{cold}}$$

While Carnot's focus was on steam engines, it turns out that his ideas had profound universal significance. Understanding his description of the energy cycle driving steam engines also provides us with a key to understanding the Cosmos, the Earth's climate, and many types of extreme weather events. Figure 3.1 shows a diagram of a Carnot Cycle. On the left and right we find reservoirs filled with hot and cold

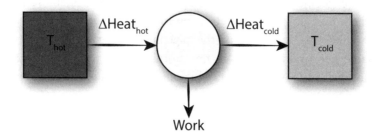

Figure 3.1 A schematic diagram of Carnot's Heat Engine.

fluids. Changes in heat energy (ΔHeat) will pass through a mechanical device (the circle), which will convert this exchange of energy into work. According to the first law of thermodynamics, in a closed system energy must be conserved. So the exchanges of heat and work must sum to zero. According to Carnot, however, the efficiency of this engine, its ability to do work, will be proportional to the difference between T_{hot} and T_{cold}. So stronger temperature gradients are more efficient.

Unfortunately, Carnot's work received little attention, and after serving briefly in the Napoleonic Army, he developed dementia and entered a private asylum at the age of thirty-six. His ideas, however, were elaborated in 1865 by Rudolf Julius Emanuel Clausius, a professor at the Swiss Federal Institute of Technology in Zurich. Clausius coined the term "entropy," seeking a term similar to energy but based on the Greek term τροπη, which relates to transformation, turning, and change. Clausius and the British physicist Lord Kelvin placed entropy in a thermodynamic framework, thereby adding an important new law of physics. While Isaac Newton was right, energy had to be conserved, entropy in a closed system also had to increase. Because heat energy will never spontaneously flow from a cold to a warm body, energy gradients in a closed system tend to diminish over time. Entropy therefore increases, and this process is irreversible, giving rise to the "arrow of time," a term coined by Arthur Eddington.

Entropy can be defined as an inverse measure of a system's *available energy* over time; alternatively, it is a measure of a system's disorder. Often, *available energy* is associated with the gradients within a systems' molecules. Pour a very strong acid and base into water and pow: you will get a strong exothermic (heat-releasing) reaction but (bubble-bubble-stir) will likely end up with something similar to salty water. Entropy increases. The acid-base gradient disappears. Pour hot and cold water into a reservoir: motions will arise as dense areas produce flows toward less dense regions, leading to an equilibration of temperature. We can use these gradients to do work, like drive a steam engine. But when we drive the turbine, there will be mixing. The very hot will get a little cooler, the very cold a little warmer. The gradient diminished, our heat engine will eventually become less efficient, entropy ever increasing as our ability to do work diminishes.

Think of pouring cream in your coffee. Swirl, look, ooh – pretty pattern. But wait long enough, and everything is perfectly mixed. The party must wind down.

Think of dropping a very hot slug of molten lead in your coffee cup and then wrapping it in an infinitely large Styrofoam wrapper (no energy goes in, no energy comes out). At first, when there is a very strong temperature gradient, lots of coffee movement will be going on. Over time, though, the temperature gradients will diminish. Since the temperature gradients drive motions of the liquid, these will cease. No energy escapes, yet the increase of entropy brings the system to "heat death." All energetically closed systems face this inevitable destiny.

The good thing about these thought experiments is that they are just that: experimental. In other words: **constructed and hypothetical**; and in other words: often **wrong**, at least locally.

To understand the bright side of empty, to better appreciate the life-shattering beauty emerging from the twist of wave, wind, branch, and wing, and maybe preserve this planet for our children, we need to understand entropy and its antonym: **negentropy**.

And that we are fundamentally living in an energetically **open system**.

And that this life-supporting, energetically open system has arisen precisely because gravity has produced the Big Empty.

Gravity creates gradients of energy and matter. It pulls in matter, and these local increases in mass density inculcate positive feedbacks, pulling in more matter. Soon enough, no matter what, we get frighteningly high densities and potentially life-sustaining nuclear fusion. Gravity pulls in matter, forming and fueling stars, which can support life. To get to this point, however, requires the Big Empty – the vast swaths of empty space. Gravity is a Big Empty generator: you can't have pockets of intense density without much larger areas of relative emptiness. The universe's "empty prairies" are inextricably linked to the bustle of her dense "big cities." Over time, **entropy can decrease in a local place** (like our solar system). This means that order and complexity, over time, can increase locally. Life might evolve. But, the local coalescence of matter must be offset by an increase in entropy (and a decrease in matter and gravity) somewhere else. Decreasing entropy means increasing negentropy, increasing complexity, and an opportunity for the evolution of life. The term 'negentropy' refers to a tendency for a localized system to become more ordered and complex over time.

So to understand the "value" of the universe we don't divide stuff by volume, or the number of kind or brilliant communications by kilometer squared. Focus on the bright side of empty, and the fact that stars can support life – and the universe is a wonderful source of innovation and attraction.

Understanding energetically open and closed systems can help resolve the central question: If entropy must increase, then how could life be possible, much less possibly probable? In the question resides the answer: embedded in the Big Empty.

A lot of empty may be the best thing that ever happened to us. If it takes a village to raise a child, it takes a mega-load of nothing to offset the entropy-negating existence of a star. Think of it this way: What are the chances that a bunch of hydrogen, helium, and other elementary molecules would just self-assemble to become a star?[2] *Nada*. But the natural tendency to

---

[2] Fred Adam, *Origins of Existence: How Life Emerged in the Universe* (Simon and Schuster, 2010).

47 / The Earth Is a Negentropic System, or "the Bright Side of Empty"

reestablish a randomized spatial distribution is offset by the force of gravity. I am totally pro-gravity. Gravity rocks hard. Gravity is our greatest friend in the fight against boring. Gravity makes stuff happen.

At a universal scale, the "empty" and "cold" of the universe are requisite components of the islands of heat and complexity we love so much. The universe is not "hostile" to humans, but we are still terribly fragile, and the scale of the cosmos grand almost beyond our imaginings. The emptiness of the universe, from this perspective, is like the silence between the notes that brings forth music.

Outside our windows sunlight streams down. Climate scientists study where that energy goes, what it does, how it moves through the Earth system. As Chapter 2 explored, the differential heating of the poles and equatorial regions drives the Hadley Circulation, forcing winds to converge near the equator, and then returning poleward aloft in the upper troposphere. These upper-level flows radiate energy and cool, sinking in the subtropics, giving birth to the deserts and semi-arid areas near the Tropics of Cancer and Capricorn. This persistent pattern avoids the curse of entropy by relying on imports of cheap energy, from the sun, which are ultimately based on imports of matter from the distant, now empty, parts of our galaxy that ended up in our star.

So ultimately, history, consciousness, and evolution appear to depend on the quite specific characteristic of an inside and an out. Energetically closed systems, in which there is no outside, will inevitably experience increases in entropy and succumb to heat death. Gradients will decay, available energy will eventually disappear. Energetically open systems – those with a within and a without (like Earth) can keep entropy at bay while complex systems continue to evolve. Without the vast emptiness of space, randomly intermixed matter would produce a sterile form of heat death. Thanks to our Earth's ability to maintain a delicate radiative balance, explored more fully in the next chapter, our complex climate supports all the wonders of history and ecology. There are common aspects to entropy-eating negentropic miracle machines. Systems like our galaxies, nebulae, our planet, ancient stromatolites (Figure 2.6), and our bodies. These complex systems all share a common means of "informing" – producing a within and a without – and evolve through time within ranges of energy gradients that support the evolution of semi-ordered behavior. Climate change is threatening to push the Earth outside our current life-supporting 'Goldilocks' zone. We all need to understand the basic physics of miracle machines. So ...

Principle 2: Energy Gradients Make Stuff Happen – or Not

... With my last gasps of air, with a flickering candle illuminating this my last will and manuscript, I would describe to you how our Earth functions as a fragile flame" The process is beautiful. We begin, with a lot of void, and a very little something (the Sun). The sun imports disorder (entropy) from the empty reaches of space. Extreme pressure gradients support fusion, which emits high-energy "shortwave" radiation. Visible and ultraviolet light are examples of shortwave radiation. This radiation is absorbed by the Earth, and re-emitted as thermal infrared "longwave" radiation. This is the energy you can feel when you put your hand over a very hot parking lot. Longwave radiation has (surprise) longer wavelengths and lower frequencies than does shortwave radiation. Very hot objects (like the Sun) emit mostly shortwave radiation. Not so hot objects (like the Earth) emit mostly longwave radiation. Greenhouse gasses trap the emitted longwave radiation, potentially creating an energy imbalance.

Luckily for us, the sustained supply of energy from the Sun has supported a continuous, fairly stable flow of energy for billions of years. Because the Earth system is energetically open, constantly receiving energy from the Sun, life and complexity can evolve. During the seventeenth century, Isaac Newton made one of the most tremendous leaps of human understanding. He linked terrestrial and celestial gravity. As above, so below. As without, so within. As fell the apple, so the Moon orbits the

Earth. Great science exposes connections. Modern science depends more on group effort, but our collective advances are no less stunning. In lieu of Galileo's telescope, we have a digital imaging revolution rapidly revealing wonders at all spatial scales. As scientists map, model, and sometimes predict the evolution of these systems, we find common laws that operate at all scales. To understand climate change we need to understand the basics of these laws. One of the most important sets of laws is thermodynamics, how gradients of heat (thermos) relate to changes in direction and velocity (dynamics).

One of the main laws of thermodynamics tells us that within an enclosed system, entropy must increase over time. As the organization of the molecules within the system becomes more disordered and random, the amount of energy available in the system to do work will decrease. Figure 3.2 shows schematically such an evolution toward greater entropy. Warm molecules are denoted by white symbols; cold molecules are denoted by dark gray symbols. At Time 1, there is a clear temperature gradient – eight warm molecules on the left – and the resulting pressure differential produces a well-defined pattern of motion (indicated with arrows). At Times 2 and 3, the temperature gradient has diminished, with only six and then five warm molecules on the left-hand side. The corresponding pressure gradients and associated motions become muted and disorganized. By Time 4 we are approaching maximum entropy. The substance is now well mixed, with equal numbers of warm and cold molecules on the left- and right-hand sides. Small random motion and mixing will continue, but the spontaneous organization of coherent large-scale structures is statistically impossible.

Luckily for us, the Earth is far from energetically closed. The Sun supplies a constant source of high-quality energy, and this energy supports and maintains gradients of many kinds: temperature and pressure gradients that drive our atmospheric winds and oceanic currents, chemical gradients that help support the evolving complexity of life.

What might happen in such an energetically open system? Figure 3.3 shows how a continuous input of solar radiation, over

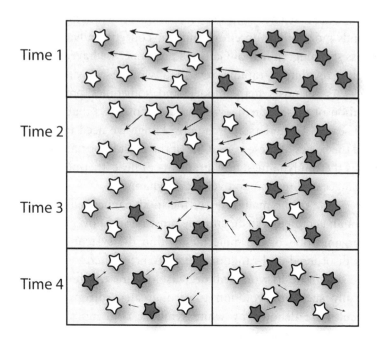

Figure 3.2 A closed system experiences heat death. White and grey stars denote warm and cold molecules. Arrows depict motion. Time 1 describes a system with a strong heat gradient, with grey stars warmer than white stars. A well defined pressure gradient produces coherent large-scale motions. In Time 2 and Time 3 mixing between the warm and cold molecules results in more chaotic pressure gradients and more turbulent motions. By Time 4 the system has experienced 'heat death'; the distribution of warm and cold molecules is completely random, all temperature and pressure gradients are very localized, and all motions are small and incoherent.

time, can support the Spiral of Life.[3] Electromagnetic radiation comes in different flavors, or wavelengths. Cold objects emit

---

[3] Interested readers may like "The Tree, the Spiral and the Web of Life: A Visual Exploration of Biological Evolution for Public Murals," www.mitpressjournals.org/doi/pdf/10.1162/LEON_a_00321. Joana Ricou and John Archie Pollock discuss their development of the "Spiral of Life" as an accessible symbol for the evolution of life – a representation that avoids the shortcomings of outdated "Tree of Life" presentations.

51 / The Earth Is a Negentropic System, or "the Bright Side of Empty"

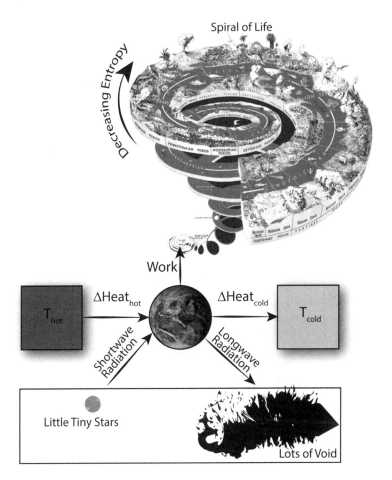

Figure 3.3 The Earth–Sun Miracle Machine drives the spiral of life. Spiral of geologic time produced by Joseph Graham, William Newman, and John Stacy, 2008, "The geologic time spiral – A path to the past." pubs.usgs.gov/gip/2008/58/

low-energy longwave radiation. Hot objects emit high-energy shortwave radiation. The sun emits shortwave radiation, primarily in the visible-light portion of the electromagnetic energy spectrum. This energy is absorbed and re-emitted by the colder Earth as thermal infrared longwave radiation. Because the Earth is energetically open (i.e., always getting energy for free), the Earth system avoids a local increase in entropy and heat death

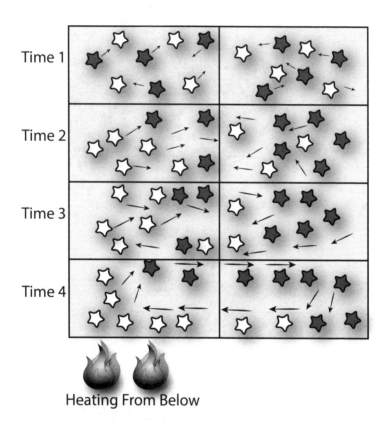

Figure 3.4 An energetically open system is heated from below. White and grey stars denote warm and cold molecules. Arrows depict motion.

(Figure 3.2). Sustained over billions of years, this negentropic behavior supports temperature gradients that drive various circulations at scales ranging from global climate to microscopic and chemical processes.

Understanding the link between differential heating and circulations provides deep insights into our long-term average climate, many anomalous climate and weather extremes, and how climate change is already altering some of these patterns. In general, the relationship between temperature gradients and circulations in the atmosphere and oceans is relatively easy to understand. Figure 3.4 shows a schematic representation of the development of a sustained circulation. At Time 1, we begin at Time 4 in Figure 3.2. The molecules are randomly mixed, the system has reached maximum entropy, and motions are small and random. Now we import energy, heating the bottom left section. This figure is generalizing the way that an external heat source can maintain temperature gradients and drive atmospheric circulations. In the real world this differential heating occurs in all directions: equator to poles, east to west, and up and down. From the equator to poles, the Sun preferentially heats the tropics, driving the equator-to-pole Hadley Circulation. From east to west, preferential heating near Indonesia and the Congo and Amazon Basins drives the Earth's east-to-west Walker Circulation. And vertically, preferential heating near the surface destabilizes the atmosphere, helping fuel thunderstorms and cyclones.

In Chapter 2 we have already discussed circulations similar to that in Figure 3.4. Figure 3.5 summarizes some of these patterns. If Figures 3.4 and 3.5A are interpreted as preferential surface heating near the equator, and the vertical dimension corresponds to height in the atmosphere, the north–south differential in solar heating between the tropics and the poles provides available potential energy that drives the Hadley circulation (Figure 2.4).

If Figures 3.4 and 3.5B are interpreted as an east–west up–down atmospheric transect across the equatorial Pacific atmosphere, then these schema describe how preferential warming in the western Pacific and over equatorial South America and Africa maintains east–west temperature gradients that provide the available energy driving the Walker Circulation (Figure 2.5).

In the vertical dimension, tropical meteorologists often use the term "Convectively Available Potential Energy" (CAPE) to describe the amount of energy available to produce severe storms or hurricanes. To understand CAPE, think about the *relative* energy contained within a parcel of air that was near the surface of a warm tropical ocean but just got shoved upward. The *difference* between the parcel's temperature and the surrounding atmosphere is directly related to the parcel's

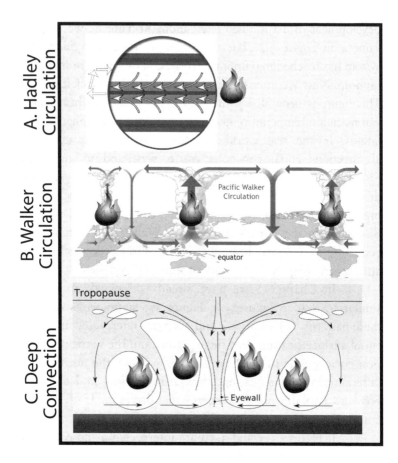

Figure 3.5 Heat gradients produce various important atmospheric circulations. North–south heating differences drive the equator-to-pole Hadley Circulation (A). East–west heating differences along the equator support the east–west Walker Circulation (B). At smaller scales, intense vertical and horizontal heat releases fuel tropical storms and cyclones (C).

"buoyancy," or tendency to keep rising. A hot air balloon rises because it is that: a **hot** air balloon. Hot here is only interesting in a relative sense. It rises because it is hotter (= less dense) than the surrounding air. CAPE also goes up when air becomes moist, and the CAPE calculation takes into account the "heat" associated with evaporated water vapor (a point we will return too). This can be very important when trying to understand climate extremes, including hurricanes and floods.

Here is a description of such an event as a (temporarily) negentropic process. First, there is a fluctuation, say a chaotic wind squall, like an atmospheric "rogue wave" that flaps the butterfly's wing, creating a temporary convergence of moisture and CAPE. Most waves, be they sound, light, or watery, tend to add linearly: i.e., when wave "a" encounters wave "b," the resulting pattern is just wave "a" plus wave "b." Sometimes such natural meanderings of our atmosphere bring together squalls, which in warm, moist environments produce uplift and condensation.

Meandering wave fluctuations can trigger the release of Convectively Available Potential Energy. When a warm parcel of air is lifted up into a cooler environment, higher up in the atmosphere, this produces available potential energy, because the parcel is buoyant relative to the surrounding air. The parcel will cool as it ascends, and when it cools enough, water vapor contained in the parcel will start to condense. When condensation begins, that also increases the available potential energy, because the "latent" energy contained in the parcel's atmospheric water vapor is released. This also makes the parcel warmer than the surrounding air.

Now we may enter into the land of positive feedback, where things can get interesting. We all know that nature abhors a vacuum. So when parcels of air start to ascend, they leave behind a region of low pressure, which helps to pull in air from the surrounding region. If this air is moist and warm, and converging (i.e., coming in from multiple directions), we have several positive feedbacks going on. First, just in terms of conservation of mass, when air masses converge near the surface, this tends to produce vertical motion. Since they can't go down, two colliding air masses tend to go up. Furthermore, if we have converging moist warm air, we have converging CAPE. These processes can act to reinforce the strength of our initial convective outburst.

Once warm moist air starts rising, it will often keep rocketing on up through the colder portions of our atmosphere until it reaches a "temperature inversion," which can often be the "tropopause" about 12–15 kilometers above the Earth's surface. At the tropopause, the atmosphere reaches a very cold −55°C. Above the tropopause we find an increase in atmospheric ozone, which absorbs sunlight. So, above the tropopause temperatures increase, which 'inverts' the typical tendency for air temperatures to decrease with height. This inversion can form a lid, capping the vertical ascent of clouds – like the common "pancakes" we often see topping massive cumulonimbus clouds.

At the upper levels of the atmosphere air parcels move away from cloud tops, cool, and sink. So tropical storms often form mesoscale (medium scale) convective systems, comprised of relatively small convective cells associated with heavy precipitation and ascent and relatively broad regions associated with descending air. Sometimes, over the tropical oceans, storms with enough rotational spin and CAPE can turn into hurricanes, cyclones, or typhoons. These are just different names for the same phenomena. Hurricanes appear in the Atlantic and northeast Pacific. Typhoons occur in the northwest Pacific. The Indian Ocean and South Pacific host cyclones.

Figure 3.5C[4] shows a typical hurricane structure. Just like a thunderstorm, a hurricane has convergent inflow near the surface and divergent outflows near the tropopause. The energy driving this circulation is provided by moist warm air. This energy ultimately comes from the ocean, primarily via the latent energy contained in evaporated water vapor and secondarily through the heating of the lower troposphere. The latent heat released in the hurricane's vertical walls is offset by the radiative cooling of air in the upper portions of the system, near the tropopause. It is the combination of this massive energy release and the hurricane's rotational structure, caused by a conservation of angular momentum, that supports the development of a spectacular and potentially devastating weather system.

---

[4] upload.wikimedia.org/wikipedia/commons/f/fb/Hurricane_profile.svg.

Conclusion

Thunderstorms, hurricanes, roses, Earth, you: these are all examples of negentropic systems. We can't cheat the universe. Matter energy and entropy must all be conserved. But the bright side of empty is that by extracting energy from one place and collecting it in another the universe is capable of supporting islands of complexity and the evolution of life.

The laws governing these systems produce beauty – beauty that I simply like to look at. Here, on this planet, we know that these complex systems have learned to do amazing things, like both persist *and* evolve. It is doing both that's the tricky part. At a molecular level, some time between 4 and 3.5 billion years ago, simple molecules such as carbon dioxide, water, and phosphate combined abiotically to form complex but nonliving organic compounds such as amino acids, nucleotides, and proteins. Then, in ways we don't fully understand, "life" arose because a complex chemical soup – maybe in a shallow pool, maybe near a deep sea thermal vent – began to develop an inside and an out, as well as the ability to harvest energy from the surrounding environment, just like a star. Perhaps ribonucleic acid (RNA) played the pivotal role, acting as both catalyst and information store. Or perhaps the origin of metabolic pathways resulted in coherent, replicating, self-organizing membrane-bound structures. Then came some means of procreation, patterning, and our first chemical "memories" arose. Somehow we arrived at a situation in which nucleic acids could both guide the creation of complex proteins and their own replication. We are uncertain of the how, but we know for sure that by 3.5 billion years ago, simple prokaryotic life arose – because we can still hold the evidence in our hands, in the form of "stromatolite" rock formations. That's pretty amazing. Just like most of the cells in our bodies, the bones of the earth itself succumb to the ravages of time, being submerged, weathered, and melted over time. The oldest rock, in fact, is only slightly older than the oldest stromatolite (approximately 4 billion years) – that is, if 300 million years qualifies as "slight."

Simple prokaryotic cyanobacteria (blue-green algae) evolve and begin the process of photosynthesis. Prokaryotes lack membrane-bound nuclei and mitochondria. Cyanobacteria played a critical role in the history of the earth by releasing (via photosynthesis) vast quantities of oxygen into our atmosphere. In this oxygen-rich environment, more complex eukaryotic cells formed, with internal organelles and nuclei holding the cell's genetic material. This led to really popular life activities such as breathing and sex. Then, perhaps due to lucky symbiotic combination of a eukaryote and a proto-bacteria,[5] mitochondria formed, providing a vital source of energy for eukaryotic life – adenine triphosphate (ATP). This process of aerobic respiration combines oxygen with sugars to produce cell fuel, which is ATP. Now the energy of the sun can be captured, stored, and transported by our bodies. This fueled an explosion of life that continues apace today.

Finishing this chapter, a deep-seated mental sigh escapes. I have written the piece on negentropy I have been pondering for years. As you look up at the stars, my hope is that you no longer see the Big Empty as negative space. Just like silence, this nothingness provides the structure for the music of the spheres.

Between my fateful brush with bee-borne death and today, I have changed a lot, and raised a family as well. I am grateful to be so different from the whiskey-drinking yahoo who almost died in the woods twenty-five years ago. I'm a lot happier now, hopefully worrying more about others than myself. It's a lot more rewarding. Valuing what we have is the first step in not letting it slip away.

---

[5] Margulis, Lynn, and Sagan, Dorion (1986). *Origins of Sex. Three Billion Years of Genetic Recombination.* New Haven: Yale University Press. pp. 69–71, 87.

4 DO-IT-YOURSELF CLIMATE CHANGE SCIENCE

Some aspects of climate change are complex and hard to fathom. Some are fairly straightforward concepts and facts that everyone really needs to understand. This book is mostly about the latter. Some of the most important mechanisms of climate change can be understood by everyone: Why do greenhouse gasses have such a direct warming effect on our planet? How does this warming intensify the impact of droughts and fires? How can this same atmospheric warming, paradoxically, also increase the frequency of extreme precipitation events and floods?

This book approaches these questions with a Do-It-Yourself (DIY) attitude. While we will occasionally make use of computer simulations results produced by some of the most sophisticated numerical models ever produced by humans, we will more frequently look at publically available data (from web sites you can access yourself) and examine these results using plain old common sense mingled with a little atmospheric science. Many aspects of climate change are as certain as gravity. These we will emphasize.

Take for example the relationship between California fires, warmer air temperatures, and drier vegetation.

For personal context, we begin on December 5, 2017. This e-mail from my friend Libby came in at just after midnight:

Figure 4.1 Firefighters battled the Thomas Fire in Ventura, California early Tuesday, December 5, 2017. Tens of thousands of people in Southern California were evacuated as wildfires raged. This photo was taken at roughly 2 AM on Manzanita Court in the hills of Ventura. The firefighters were from Los Angeles Fire Department Station 10. The single engine arrived and saved all other homes on this cul-de-sac except the two pictured. Image credit: Ryan Cullom/RLC Photography

"My town is kind of a lot on fire" (Figure 4.1). From her apartment in Ventura, Libby watched the hills explode into flames reaching six stories into the sky, fanned by 50 mile/hour (81 km/hour) Santa Anna winds. Only contained by January 12, 2018, the Thomas Fire briefly became the biggest and most damaging fire in California history, destroying more than 280,000 acres and a thousand structures. Yet this baleful record was not to last long, with the 2018 Mendocino Complex Fire covering more than 459,000 acres, and the deadly Camp and Woolsey fires in November 2018 claiming more than 80 lives and together destroying more than 80,000 structures and burning over 250,000 acres or 101,00 hectares. As I update this chapter in early November 2019, a series of powerful Santa

Anna wind events, fiercely hot and dry, funneled down the mountains and toward the Pacific Ocean, and have just subsided, leaving behind fire scars up and down the state, from the fabled hills near Hollywood to Sonoma County north of San Francisco.

The Santa Anna wind conditions complicit in California fires are driven by a strong pressure gradient between the terrestrial interior of the western United States and the Pacific Ocean; a high-pressure cell located over the western United States drives winds toward the relatively low-pressure areas to the west – the Pacific coast.[1] Climate change may be making these winds stronger, but it is hard to be sure. One recent study by Janin Guzman-Morales and Alexander Gershunov of the Scripps Institution of Oceanography, University of California, suggests that climate change may actually be reducing the frequency of these extreme wind events, especially during the early part of the rainy season. This change, together with anticipated changes in the winter rains, is expected to shift the fire season later in the year. One aspect, however, is quite clear. California, along with many other regions of the planet, has become a lot warmer. Warmer weather has reduced relative humidity values, drying out vegetation and helping extend California's fire season into October, November, and December, when a seasonal increase in winds has helped lead to many of the most deadly recent conflagrations.

In today's web-enabled world, you don't have to just take this as a matter of faith. My life is all about weather and climate data. I love it, live it, produce it, and use it to identify disasters and help people. This book will present a lot of data, and most of it you can access yourself using online tools. For example, good climate records for the United States go all the way back to the late 1890s, and we can access them using

---

[1] Guzman-Morales, Janin, and Alexander Gershunov, "Climate change suppresses Santa Ana winds of Southern California and sharpens their seasonality." *Geophysical Research Letters* 46.5 (2019): 2772–2780.agupubs.onlinelibrary .wiley.com/doi/pdf/10.1029/2018GL080261.

62 / Drought, Flood, Fire

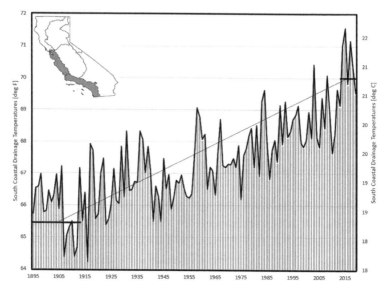

Figure 4.2 Observed June–October average air temperatures from the southern and central California coastal drainage regions. Data obtained from www.esrl.noaa.gov/psd/cgi-bin/data/timeseries/timeseries1.pl on November 8, 2019

resources provided by the National Oceanic and Atmospheric Administration's Physical Sciences Laboratory.[2] For example, I have selected the June–October air temperature data for the central and south coast regions of California (where I live). This is the period of time between one rainy season and the next, when California vegetation dries out. These dry conditions, combined with Santa Anna winds, can produce optimal conditions for fires along California's central and southern coastal drainages. How much warming do we see in the data? A lot.

Figure 4.2 shows the annual averages in degrees Fahrenheit and Celsius. Two horizontal bars in Figure 4.2 denote conditions typical in the early twentieth century (1895–1914), and over the recent past (2014–2019). We find a large degree (pun intended) of warming. Between the beginning of the

---

[2] www.esrl.noaa.gov/psd/cgi-bin/data/timeseries/timeseries1.pl.

twentieth century and today we see 4.5°F (2.5°C) of warming. Coastal California has already experienced a very large temperature increase. You can see for yourself that this long-term warming is very large compared to the typical year-to-year fluctuations in air temperatures. Typical values for these fluctuations are 1.1°F (0.6°C). Later chapters will describe how this warming contributes to drier vegetation and larger wildfires. In this chapter we will use a multitude of data sets to describe how the Earth is warming globally, and also describe how and why that warming is happening.

But before going forward, let's first ask ourselves whether the observed warming is likely to continue. To address this question, we can use the "Climate Explorer" web tool[3] to download the average of a large number of climate change simulations (described more completely later in this book) for our region of interest, the Californian coastal region. Figure 4.3 shows a time series of these modeled air temperature values, along with our observed air temperature data. The observations are the same ones shown in Figure 4.2. Both of the simulated and observed temperatures are presented as running 10-year averages to represent the long-term tendency. The last value in the black line shown in Figure 4.3 corresponds to the average observed June–October air temperature over the 2010–2019 time period. The next-to-last value corresponds to 2009–2018, and so on. The dashed line in this figure shows the predictions from the climate change models, which track *very* closely with the observations. A statistical comparison (a regression, not shown) tells us that the model explains 80 percent of the variability of the observed air temperature variations. So to date, the best estimate provided by climate scientists has performed very well. There have been some fluctuations above and below the climate change prediction, but these are to be expected, given natural climate variability.

Note the big increase in temperatures that occurred over the 2015–2019 time period. This more than 1°C jump has been

---

[3] The KNMI web tool is a tremendous resource: climexp.knmi.nl/selectfield_cmip5.cgi.

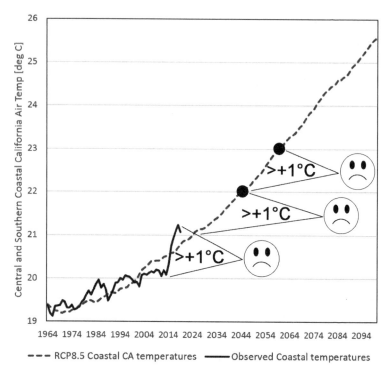

Figure 4.3 Observed and simulated coastal California June–October air temperatures. Data obtained from climexp.knmi.nl/selectfield_cmip5.cgi

annotated with a sad face. As discussed further in Chapter 11, that sad face represents a deadly and costly increase in fire extent and tree mortality. By looking carefully at both weather data and disaster statistics, we can see with our own eyes that a 1°C warming is already having dire consequences. We will see similar results over and over again, both in this chapter and in this book: a very rapid 2015–2019 warming associated with the increased frequency and severity of droughts, floods, and fires.

Beyond 2020, the dashed line in Figure 4.3 represents where temperatures are likely to trend if we continue our current emissions. By about 2045, Californians will experience another 1°C (1.8°F) of warming, and then in just fifteen more years yet

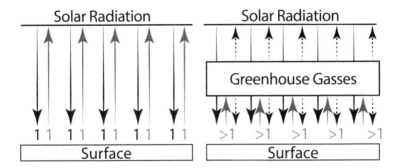

Figure 4.4 Schematic diagram showing an atmosphere with and without greenhouse gasses.

another 1°C (1.8°F) of warming. These increases are also annotated with sad faces in Figure 4.3. The intent here is serious. California just experienced a large increase in fire extent, due in part to an exceptional increase in temperatures. We will likely see this happen again, and again, over the next forty years.

Curbing emissions now will help prevent these dangerous increases. But why should you believe me? First, and perhaps most importantly, because you understand the basic physics. They are not really hard to understand.

How greenhouse gasses warm the atmosphere is pretty straightforward. It is *really* important that we all understand this. But almost no one does, because it is not something that we are taught in school. Even most people who believe in climate change don't really understand the basic mechanism of why adding greenhouse gasses to the atmosphere *has* to increase the amount of energy reaching the surface of the Earth.[4] This mechanism is all about the contrast between a "transparent" atmosphere with no greenhouse effect (left side of Figure 4.4), and an "opaque" atmosphere with a greenhouse gas effect (right side of Figure 4.4). The terms "transparent" and "opaque" refer to how infrared radiation (emitted heat energy) responds.

---

[4] I owe a debt of gratitude to my PhD advisor, Professor Joel Michaelsen, for teaching this to me in graduate school. Thanks, Joel.

Under a transparent scenario, incoming solar radiation comes down and the same amount of infrared radiation is emitted upward. This equality is indicated by the black and gray 1's on the left of Figure 4.4. The same amount of energy arrives (downward black arrows) and departs (upward gray arrows). The "flavor" (frequency) of the radiation, however, shifts. When any object emits radiation, the frequency of the emitted energy will be determined by the temperature of the object. Warmer objects will emit higher-frequency energy. Because the Sun is very hot, it emits energy with high frequencies. So, on the left of Figure 4.4 the incoming solar radiation (downward black arrows) has a high frequency, mostly corresponding to visible wavelengths (e.g. red green blue), which our eyes have evolved to detect. The Earth's surface and everything on it (like you and me), emits longer-wavelength, lower-frequency infrared radiation. In an atmosphere without greenhouse gasses (left side of Figure 4.4), there is a simple balance between the downwelling shortwave radiation (black arrows) and upwelling longwave radiation (gray arrows).

Now, let's add carbon dioxide, methane, ozone, and other greenhouse gasses to the atmosphere (right side, Figure 4.4). Furthermore, to keep things simple, let us assume that all the upwelling longwave radiation is absorbed by a middle layer of greenhouse gasses. Now, something *really important* happens. These gasses absorb the upwelling infrared energy from the Earth and re-emit it in all directions. Ironically, this symmetric emission pattern produces asymmetric greenhouse gas warming effects. On the way down, visible light passes innocently past greenhouse gas atoms because visible light lacks the right frequency to excite greenhouse gas electrons. Then this visible light energy is absorbed and re-emitted by the Earth's surface. But the energy that is re-emitted is lower frequency infrared radiation. This lower frequency infrared radiation is trapped by greenhouse gasses. So (in our cartoon world) it is all absorbed by the middle greenhouse layer, which sends half of its absorbed energy back down to the Earth's surface. Which gets warmer, and sends more energy back upward. This re-emission process mandates an increase in the Earth's temperature. It is straightforward physics, a process that we understand far better than gravity, and there is nothing to "believe" in. I don't want you to be concerned about climate change because you believe me. Rather, I want you to be concerned because you understand how it works.

Understanding how humans can so radically alter the atmosphere so quickly begins with understanding that the atmosphere is actually extremely thin. This is not how we perceive the atmosphere. We look outside at the sky and it looks really big. But in fact, only a relatively thin band of atmosphere separates us from space. Over our heads looms about 1 million kilograms (2.2 million pounds) of atmosphere. But the density of this layer drops exponentially, as the mass of the air below is required to sustain the mass of the air above. If we somehow managed to climb to a height of 16 km, about 90 percent of the atmosphere would lie below us. From the South Pole to the North Pole the Earth stretches about 20,000 kilometers – 1,242 times 16 km. Drive your car straight up at highway speeds, and you reach the edge of the atmosphere – about 10 miles (16 km) – in about 10 minutes. Imagine an egg painted blue. The atmosphere is as thick as the blue paint.

Another key insight revolves around the difference between the atmosphere's *volume* and its *mass*. Humans naturally tend to perceive the atmosphere as vast, but science is all about overcoming our limiting misperceptions. Consider, for example, the pre-Socratic Greek proto-scientist Heraclitus,[5] who lived in Ephesus around 500 BC. Heraclitus famously stated, "The sun is the width of a human foot." Now this *is* actually an experiment you can try at home. While Heraclitus was basically correct, as you can check by lying on your back outside, his conclusion can be quite misleading. By analogy, when we look at the sky, we perceive it as massive. But there is really not that much there up there. While the atmosphere is a very large place, the amount of mass it contains is relatively small.

[5] plato.stanford.edu/entries/heraclitus/.

We can try to wrap our minds around the true size of the atmosphere by taking all the mass of the atmosphere, prescribing a density equal to air's density at the surface, and estimating the resulting volume. This cube would be about 800 kilometers on each side.

This may still sound like a vast volume. But there are a lot human beings too. Let's put the Earth's approximately 8 billion people into the cube as well. They would each be about 400 meters from each other. Now put each person in a car. Start the engines. Assume that each car emits about five *tons* of CO_2 per year.[6] Imagine those five tons of CO_2 building up in your 400 m × 400 m × 400 m cube of air. Year after year the CO_2 lingers and accumulates.

The buildup of greenhouse gasses will trap more upwelling longwave radiation. The greenhouse effect arises from two simple, very well-understood physical effects: (1) certain gas molecules absorb and re-emit longwave radiation, and (2) energy must follow a balance, i.e., on average, what is received must be given back. The first aspect, absorption, gets a little tricky in its details. Gas molecules – whirling, shaking clouds of electrons – like all the rest of matter, are mostly void. A water molecule is small, 3×10^{-13} meters across. The sentence-ending period dot on this page might contain some 333,333,333,333 water molecules. But the nuclei of the hydrogen and oxygen atoms are much smaller in diameter, so the nucleus of a hydrogen atom and that of an oxygen atom, connected end-to-end, would span a mere 6.5×10^{-15} meters across. The cross-sectional area of the hydrogen-oxygen-hydrogen nuclei is only 0.0168 percent of the area of the water molecule, so it's just like you concluded as a teenager: a pointless empty void. Given the vanishingly miniscule chance of a light photon actually striking a nucleus, why do molecules

---

[6] The 2019 Global Carbon Project estimate for global CO_2 emissions was approximately 37×10^9 and the 2019 global population was approximately 7.8×10^9, so 2.6 tons per person would be more accurate. But the main point of this section is to try to visualize our relationship to our atmosphere.

absorb radiation, and why do different molecules preferentially absorb and emit specific wavelengths?

Figuring this out is actually why Einstein won his Nobel Prize. Color is quantized because light waves mainly interact with atoms through absorption and re-emission. Absorption and re-emission happen preferentially, based on the particular wavelength of light and the particular electron cloud configuration (or geometric shape) of a particular atom or molecule. What Einstein described was the link between the different energy states of electrons and the ability of atoms to absorb certain types of radiation, i.e., radiation of specific wavelengths. As you may remember from chemistry class, electrons have configurations in different "orbitals," mostly determined by the number of protons (since the number of protons and electrons almost always match). But every once in a while, an electron can absorb a specific amount of energy, and the "excited" electron will jump to the next highest orbital (energy level).

When the excited electron descends back to its old orbital level, it emits a photon with exactly the same amount of energy that it received, and with the same wavelength as the energy it absorbed. This is where the change in wavelength between the incoming "shortwave" radiation from the sun and the outgoing "longwave" radiation from the Earth's surface becomes important. Shortwave radiation (mostly visible light) does not "see" water vapor or carbon dioxide. Green photons pass through water vapor molecules without interacting with the water vapor electrons. The Earth absorbs this visible light. The Earth re-emits this energy as "longwave" infrared radiation.

Infrared radiation, it turns out, is able to "see" greenhouse gasses. These atmospheric gasses tend to absorb and re-emit the Earth's longwave radiation. But when they do so, they have to emit in all directions. And this makes all the difference.

The parable of Stinky, Bif, Dinxie, and Moo will help us understand why (Figure 4.5). Stinky and Bif are dogs, and Dinxie and Moo are cats, but their inter-species differences do

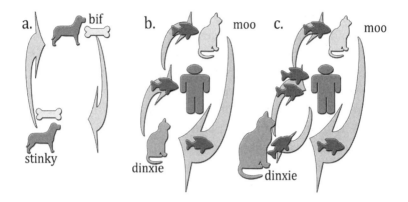

Figure 4.5 Cats and dogs in love. A story of passion, greed, and education in which the greenhouse effect is revealed, and Dinxie's weight loss program is derailed.

not affect the underlying physics. Stinky and Dinxie represent the Earth's surface. Bif and Moo represent the top of the Earth's atmosphere. Stinky and Bif are in love. Theirs is a simple world (Figure 4.5a). Bif receives a bone from the solar energy god.[7] Bif gives Stinky the bone. Stinky gives Bif the bone. Bif can't catch the bone,[8] but that's okay, because the sun god gives Bif another bone, and the cycle repeats. Stinky's final temperature will be matched to the energy he received from Bif, so that he is emitting exactly the same amount of energy he received.

Now, let's look at love in a little more complicated world. This world's atmosphere has a middle layer, home to a dubious complicator we'll call Taxem. Taxem is so chock-a-block full up with water vapor, carbon dioxide, and methane, he smells like a Kansas high plains cattle feedlot crossed with the Los Angeles 405 Freeway. As our story begins (Figure 4.5b), Moo also receives an energy fish from the god of the sun. Meoww! He gives his fish to his best friend Dinxie. Because Dinxie is a cool cat, she receives a "shortfish" but emits a "longfish," and is unselfishly giving it back ... when whoosh – Taxem steps in

---

[7] His name is Larry. But that's not important. Really.
[8] Why can't Bif catch Stinky's bone? A little riddle to see if you've been paying attention.

and eats that fishy up. Taxem does this because his methane and water vapor and carbon dioxide preferentially absorb longfish. In this simple world (Figure 4.5b), we'll pretend that Taxem can re-emit Dinxie's longfish in a preferential way (up). In this case, we end up with a final situation very similar to the Tale of Stinky and Bif. Taxem gives Moo a fish. Moo can't catch the fish, but that's okay, because the sun god gives Moo another fish, and the cycle repeats. Again, Dinxie's final temperature will be matched to the energy she received from Moo.

Let's look at cat love in an even more complicated world (Figure 4.5c). In this world, Taxem operates more like our real world, emitting energy both up and down. Moo begins by giving a fish to his best friend Dinxie. Who unselfishly is giving it back ... when whoosh – Taxem steps in and eats that fishy up. This time, though, Taxem has to give the same amount of fish upward as he does downward. How can everyone be happy?

We can solve this simple energy balance problem using algebra. It really does come in handy sometimes, and what follows may be one of the most important pieces of science that few people know. We are describing something that is absolutely vital to understand. We are going to work our way through a simple three-layer energy balance model (similar to Figure 4.5c). We have three atmospheric levels: the surface (Dinxie), the middle of the atmosphere (Taxem), and the top of the atmosphere (Moo). $M_\downarrow$ (Moo) represents the shortwave energy flowing into our atmosphere. $D_\uparrow$ represents the upwelling longwave radiation flux from the surface (from Dinxie). $T_\uparrow$ and $T_\downarrow$ represent the upward and downward fluxes from greenhouse gas radiation trapped in the middle of the atmosphere (from Taxem). To conserve energy at the top of the atmosphere, we know that $M_\downarrow$ must equal $T_\uparrow$ (i.e what goes out the top must equal what comes in the top). Similarly, at the surface we know that $D_\uparrow$ must equal $T_\downarrow + M_\downarrow$, i.e., the amount of energy emitted by the surface must equal the amount of energy received by the surface. We also know that Taxem must emit the same amount of energy up and down. So we have three equations:

$M_\downarrow = T_\uparrow$
$D_\uparrow = T_\downarrow + M_\downarrow$
$T_\downarrow = T_\uparrow$

We have three equations and three unknowns. We can use simple algebra to solve for $D_\uparrow$ by substitution: $D_\uparrow = T_\downarrow + M_\downarrow$. Since $M_\downarrow = T_\uparrow$, then $D_\uparrow = M_\downarrow + M_\downarrow$ and $D_\uparrow = 2M_\downarrow$.

Wow. We have just derived a simple model to describe the basic mechanism of the greenhouse effect. In the *absence* of a radiatively active middle atmosphere (Figure 4.5a), and ignoring complicating factors such as evaporation and the effects of clouds and convection, the surface temperature would be exactly warm enough to radiate enough energy to match the incoming radiation. In the *presence* of a radiatively active middle atmosphere, however, the fact that the middle atmosphere absorbs and re-emits longwave radiation ensures that the energy emitted by the Earth will be substantially *more* than the amount of incoming solar radiation. This may seem paradoxical, but we have just used a simple numerical model to explore for ourselves how this process works. The basic idea behind the greenhouse effect is simply this: energy must be balanced, but the middle atmosphere must absorb and emit radiation in all directions, while the surface of the Earth only radiates energy up. Like death and taxes, these assumptions are a sure thing. When you put these simple truths together, you get a surprising result; the surface of the Earth warms substantially. Increase the amount of greenhouse gasses in the middle atmosphere, and the surface of the Earth must get warmer. This is simply physics and logic.

While simple, the toy model we have explored here is quite similar in principle to the numerical "radiative transfer" models used by climate scientists. While some of the details can be very complicated, and very important, the core science behind radiative transfer is absolutely certain and relatively simple. The greenhouse effect is not a theory, but rather basic physics. Like gravity, it is an everyday phenomenon of our world. Unlike other types of physics, like the laws that describe the motions of falling apples, most of us have never been taught how radiative transfer works, and it is difficult for us to experience it as viscerally as gravity's ten meters per second per second acceleration.

So, it is ironic that "global warming" remains such an intense topic of "debate." The physics behind the effect of greenhouse gasses is much better understood than the mysterious mechanisms creating gravity, and much more direct than the circuitous interactions giving rise to evolution. While its effects are easy to quantify, there is still considerable discussion and debate about what actually causes gravity. At non-quantum spatial scales, Einstein's theories of relativity describe gravity as function of the curvature of space-time; more matter produces more curvature. But why? At a quantum level, considerable debate continues – with some radically differing opinions about the basic structure of reality – and the processes that account for gravity. Is gravity produced by tiny quantum gravitons? By a seven-dimensional super string structure? By m-dimensional manifolds? Despite continued efforts by some of our most brilliant minds, we really aren't sure.

Compared to gravity, the greenhouse effect is easy: physicists really understand at the core level the processes involved, and have developed effective models to represent it. What's different from gravity, however, is that our personal experience of greenhouse warming is very limited. Isaac Newton's great insight – linking gravitational attraction on Earth to the attraction of planets, the sun, and the moon – was brilliant. It took a great leap of insight to link the observable terrestrial reality of weight – the downward pull of mass on Earth with the invisible force that drew the planets toward the sun. Brilliant, mathematically elegant, but ultimately something we still don't understand very well scientifically. But on the other hand, gravity is deeply familiar to us all. When we are young, we fly over the top of our handle bars, or slide on gravel and lose the skin off our palms. We grow, stretch, and learn, laughing in the face of gravity's pull.

The challenge to understanding the greenhouse effect is to hurtle our mind's eye beyond our personal experience. To ponder and accept the grand balances of nature that enable, and encircle, the wonder of our lives.

We should also recognize that the greenhouse effect is not inherently "bad." Without the greenhouse effect, our planet would be much colder, too cold to make life comfortable for us. The greenhouse effect is just one factor among many affecting the Earth's climate. In many natural systems, "balance" is typically shorthand for complex sets of competing forces that interact to produce complex behavior. In living systems, however, this process is often formalized by monitoring networks and feedback loops capable of supporting "homeostasis." Our autonomic nervous systems, for example, might monitor our body temperatures and cause us to sweat when we grow too warm. Dealing with the impact of greenhouse gasses might not be so simple. While we have been blessed to live on a fantastic "Goldilocks Planet," we have broken the chair, we have eaten the porridge, and we have made a mess of the bed. Now the big bad bears are coming home. As we discussed in Chapter 2, the Earth's atmosphere has been "just right": unlike Mars, which has very little atmosphere, no greenhouse effect, and is too cold to support life, or Venus, which has a very thick atmosphere, very strong greenhouse effect, and is too warm for life, the Earth's temperature range has been relatively stable as humanity emerged, and the greenhouse effect helped keep our planetary temperatures in the magic range allowing frozen, liquid, and gaseous water. But when it comes to greenhouse warming, there can certainly be too much of a good thing.

Observational Evidence for Climate Change

Now, using data, let's review the primary evidence for current concerns about recent global warming.[9] Rather than ask

---

[9] Most of the data examined here was obtained from KNML climate explorer (climexp.knmi.nl), which is an archive of data and climate simulations used to support the Intergovernmental Panel on Climate Change's Fifth Assessment Report.

you to "believe" in climate change, we look at the observations and simulations at the center of the Intergovernmental Panel on Climate Change (IPCC) Fifth Assessment Report,[10] and examine the factors contributing to the rapid acceleration of emissions over the last decade.

There are three primary lines of evidence. First, greenhouse gas impacts originate from the basic physics of radiative transfer (as described above). Believing in global warming is like "believing" in gravity. The physical process is not that hard to understand, but it is not something most people have learned about. Second, a wide variety of different data sources all converge on the same story: we are experiencing a huge disruption in our climate, and this disruption will manifest itself in the twenty-first century, just as population growth and increasing demands for water, food, and resources will place unprecedented strains on our Earth's ecosystems. Third, at a global scale, the agreement between the climate model predictions and the observed changes that have already happened is very close. If the models' future predictions are right – and they almost certainly are – and we continue to emit greenhouse gasses at an accelerating rate, then twenty-first-century global warming will be dramatic, dangerous, and disruptive. Again, simple physics.

We next focus on three separate but convergent sources of observational evidence: (1) terrestrial air temperature records, (2) observations of sea surface temperatures, and (3) observations of global sea levels. We show that all these independent records converge on dramatic and rapid increases in land and ocean temperatures. We will also examine a large collection of climate change simulations, which have been produced to inform the Intergovernmental Panel on Climate Change's (IPCC) Fifth Assessment Report.

We begin our voyage of discovery simply, by plotting estimates of global land temperatures going back to 1880 (Figure 4.6). This type of data, based on thousands and thousands of monthly air temperature observations, is one of the

---

[10] www.climatechange2013.org/report/, released on September 30, 2013.

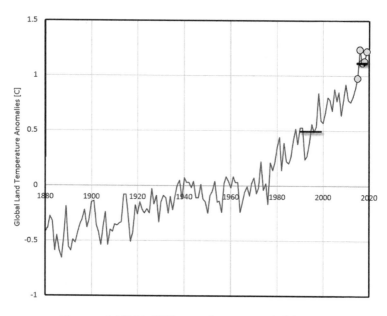

Figure 4.6 NASA GISS annual estimates of global land temperature anomalies.

main sources of evidence for climate change. These values are shown as anomalies (differences) from the 1951–1980 mean. One appealing feature of air temperature observations is their independence. Each day thousands of measurements are taken, and each location and time sample provides another piece of evidence. While systematic shifts or sources of error can occur at a weather station site (say, by moving the weather station to a lower, hotter location), it is extremely unlikely that such systematic errors could affect large numbers of stations. Time series like Figure 4.6 are based on thousands of thermometer observations, with each temperature observation providing an independent "vote." As noted by hundreds of researchers, warming on land has been dramatic, according to the available information. Furthermore, the rate of this warming is accelerating, in line with predictions from climate change simulations.

These data indicate two epochs of rapid warming. The first epoch, from roughly 1900 through 1940, is thought to be due primarily to changes in land cover and land use, solar radiation, and volcanic disruptions.[11] After the 1970s, and through 2019, a rapid epoch of warming reasserted itself, caused by the rapid accumulation of greenhouse gasses. During the monster El Niño of 2015/16, massive amounts of energy were released from the Pacific Ocean, and in 2016 and 2017 global temperatures jumped **way** beyond the historical maxima, to a 1.2°C anomalies. The years 2015, 2016, 2017, 2018, and 2019 were, by far, the warmest years in the historical record. These years are marked with circles. Lines also denote the 1990s average (+0.5°C) and the 2015–2019 average (+1.1°C). Just in the last twenty years we have seen a massive 0.6°C warming. This book examines climate extremes during these most recent very warm years (2015–2019).

Another valuable source of long-term climate information is provided by the logs of oceangoing ships. Dependent on wind and wave, sailors have always paid close attention to ocean conditions, and good ocean measurements stretch back to the late nineteenth century in many places. The temperature of the ocean can provide valuable information about a ship's location and local fishing conditions. Like terrestrial air temperature archives, these thousands of sea surface temperature observations each provide separate pieces of evidence, helping us measure and map the rate of change. Again, these estimates are based on a large number of independent temperature estimates from ships and buoys. In Figure 4.7 the global average ocean temperatures are shown as anomalies (differences) from the 1951–1980 mean. It takes an enormous amount of energy to heat just one cubic meter of seawater, so the ocean has a tremendous amount of thermal inertia. So, ocean temperatures tend to change more slowly than the land. During World War II (the 1940s) we may see some questionable jumps in the ocean temperatures, potentially related to disruptions in shipborne temperature observations. Nonetheless, we see a high degree of similarity with the air temperatures plotted in Figure 4.6, especially in the twenty-first century. While terrestrial air

---

[11] IPCC 5th assessment report, www.ipcc.ch/report/ar5/wg1/.

temperatures are more variable from year to year, the general structure of the land and ocean temperature time series are very similar to each other. While the ocean has warmed a little less than the land, because of its high thermal capacity and ability to lose heat through evaporation and vertical mixing, ocean temperature changes have tracked very closely with terrestrial air temperature variations. Both have exhibited rapid warming. The oceans have also warmed about 0.5°C between the 1990s and 2015–2019. We also see that 2015, 2016, 2017, 2018, and 2019 (circles) are way warmer than any previous observations, even when compared to 2014 (diamond).

Observations of ocean levels from tidal gauges[12] and Earth-orbiting satellites[13] provide a third independent set of global temperature observations. As water warms, it expands in a very predictable way, so slow changes in sea level heights are seen as a key indicator (and outcome) of anthropogenic climate change. Global sea levels have risen substantially (Figure 4.8). Since the deep ocean warms more slowly than the sea surface, or land, this long-time series of ocean levels lags behind global land and sea surface temperatures, but still provides further corroborating evidence that substantial warming is well under way. These rising sea levels are also associated with increased coastal flood risks, especially during cyclones and hurricanes. Massive sea swells, like those that flowed into New Orleans during Hurricane Katrina in 2005 and into New York City during Superstorm Sandy in 2012, were tremendously damaging.

Figures 4.6–4.8 show that three sources of totally independent data – terrestrial weather station observations, ship-based sea surface temperature estimates, and global sea levels – identify a large and coherent warming signal.[14] These

---

[12] Jevrejeva, S., A. Grinsted, J. C. Moore, and S. Holgate, "Nonlinear trends and multiyear cycles in sea level records." *Journal of Geophysical Research* 111 (2006), C09012, doi: 10.1029/2005JC003229.
[13] climate.nasa.gov/vital-signs/sea-level/.
[14] Lots and lots of satellite data also provides corroborating evidence over the past thirty-five years. We know the most about the Earth when most of greenhouse warming has occurred – over the past thirty-five years.

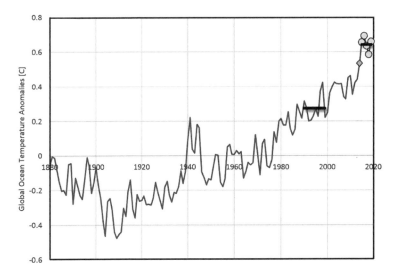

Figure 4.7 Global annual ocean temperature anomalies, based on the NOAA Extended Reconstruction dataset, version 5.

data sets are copious, consistent and continuous, stretching back to the nineteenth century. Historically, humans paid attention to temperatures and tide levels because they were important for agriculture and maritime navigation and trade. These three independent data sets tell us that the Earth is warming rapidly, and that even within our "new normal" the 2015–2019 time period has been exceptional. The data presented here do not imply that the Earth has never been warmer – it probably has been at some point in the geological past – but it does suggest that our rapid expansion in population, fuel usage, and emissions has been accompanied by a very large, rapid, and consistent warming of the Earth's atmosphere and oceans.

While the Earth may have been warmer in the past, the recent rate of warming is almost certainly unprecedented. And what we saw in the time series presented here (Figures 4.2, 4.3, 4.6, and 4.7) is that warming in the last ten years has been particularly rapid. As we will discuss in later chapters, this rapid warming is already presenting an existential threat to many organisms, including humans. The speed of these changes rule out short-term evolutionary solutions, leaving only migration or

Figure 4.8 Global sea level anomalies. The long black line is based on historic tidal gauge observations. The shorter, more recent record is based on satellite altimetry estimates provided by NASA: www.climate.nasa.gov/vital-signs/sea-level/.

adaption. But coral reefs, poor farmers, and forests have a very limited capacity to move or adapt.

Figures 4.6–4.8, however, provide little information about the *cause* of that warming. Could changes in incoming solar radiation, linked to variations in sun spots, be the cause? The data do not seem to suggest this. Figure 4.9 shows estimates of solar insolation based on satellite[15] and terrestrial[16] observations. Incoming solar radiation corresponds with the amount of solar radiation hitting the Earth. These correspond with the downward-pointing arrows from Figure 4.4.

---

[15] Ohlich, R., "Observations of irradiance variations." *Space Science Review* 94 (2000): 15–24.

[16] Lean, J., "Evolution of the Sun's spectral irradiance since the maunder minimum." *Geophysical Research Letters* 27.16 (2000): 2425–2428.

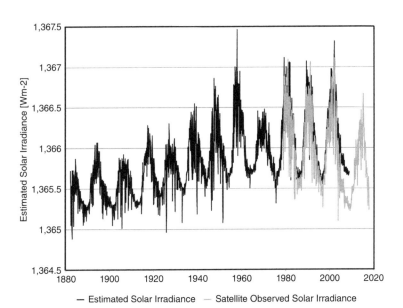

Figure 4.9 Incoming solar irradiation estimated from weather observations (black) and satellite observations (gray).

Changes in incoming solar radiation are influenced by semi-regular cycles associated with sunspots, but these variations bear little resemblance to recent variations global temperatures, suggesting that this has not been the dominant force for recent warming trends. Figures 4.6, 4.7 and 4.8 do not look like Figure 4.9. Note also that the total range of these values is only on the order of 1–2 W/m². Certainly these variations can have some influence, but they bear little similarity to the observations of air temperatures, ocean temperatures, and sea levels plotted above. We need to dig deeper.

What about results from the latest generation of global climate change models? To support climate change analyses, like the Intergovernmental Panel on Climate Change Assessment reports, international climate modeling groups from all over the world build and run state-of-the-science climate change models. Each modeling effort in this assemblage is actually a combination of models (or modules really), with each model/module dedicated to representing a piece of the Earth system. There is a model for the atmosphere, and another model for the ocean, and another model for land surface processes (like plants and rivers), and another model for the "cryosphere." Cryosphere sounds like a snow globe to me, but these critical models relate variations in temperature, precipitation, and sea surface temperatures to changes in snowpack, glaciers, and sea ice extent.

Multiple modeling centers all over the world develop their best (or several best) models. Typically, each model is initialized using land cover and greenhouse gas conditions representing the state of the world in 1850. When a simulation begins, all the variables representing atmospheric and oceanic temperatures and motions have initial values. Then they get busy perturbing each other. Winds blowing across ocean waters induce changes in sea surface temperatures. Changes in sea surface temperatures affect the patterns of winds. Changes in the winds can produces storms. Rainfall from storms can release heat in the atmosphere. This atmospheric heating can influence wind patterns. Changes in surface winds can affect sea surface temperatures. And so on.

These models are only constrained by the underlying physics of the Earth systems involved and a few external factors such as changing the amount of greenhouse gasses, the amount of particulate pollution in the atmosphere (aerosols), and the type of vegetation covering the Earth surface (farm, forest, city). At each time step, often four times a day, the models simulate the "coupled" ocean, atmosphere, and land surface. "Coupling" between the models ensures physical realism, so that energy and mass balance between the various components of the climate system. Storm systems in the simulations move east to west and west to east, tracking north and south with the passing of the sun and seasons.

To produce the Intergovernmental Panel on Climate Change Assessments, this process was carried out for many simulations for the many different models being run by the many modeling groups. While the weather on any given day, in any two different simulations, is expected (guaranteed in fact) to be very different, the *average* behavior of the simulations gives us insight into how our world is responding to human-induced changes in our atmosphere and land surface.

Slowly, between the mid-nineteenth and early twenty-first century, the external bounding conditions (greenhouse gasses, aerosols, and land cover) are changed to match the ways in which we have altered our world. Carbon dioxide concentrations change from about 287 parts per million in 1850 to about 415 parts per million in late 2020.[17] Over time, especially in the early nineteenth century, forests are replaced by cropland and human habitations.

I know from personal experience what a massive transition this could be. I grew up in northern Indiana, the flat land where soybean and corn crops stretched for thousands of square miles, filling the horizon. In the early 1990s, I was lucky to get the chance to help my grandfather put together a book describing this history (*Sod, Seed, and Sacrifice*).[18] The Funk family came to northern Indiana in the 1850s as homesteaders from Germany. To do research, Grampy and I drove around rural northern Indiana, visiting graveyards and historical societies in little towns like Fowler. There were a few pit stops at some American Legion Outposts along the way too. Grampy loved talking with people and telling stories. The story of sod, seed, and sacrifice is the story of the taming of the American prairie, whose "sod" stretched across entire states, deep into the soil, and high up into the air. Breaking that sod was a gargantuan task, often requiring massive plows pulled by a dozen oxen to cut through deep root systems developed over decades of growth. In America, corn replaced the prairies. All across the globe, forests shrank as the global population increased from 1.2 billion in 1850 to 1.6 billion in 1900. Now we can use global climate models to assess the impacts of these changes.

---

[17] www.scripps.ucsd.edu/programs/keelingcurve/.
[18] www.amazon.com/seed-sacrifice-William-Everett-Funk/dp/B0006PED5M.

84 / Drought, Flood, Fire

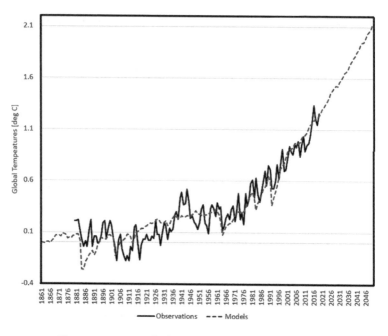

Figure 4.10 Annual observed and modeled global surface temperatures. The observed temperatures were obtained from the NASA Goddard Institute for Space Studies (GISS). The modeled time series is based on 221 climate change simulations. Future temperature projections are created using simulations from the RCP4.5 and RCP8.5 scenarios.

This data are accessible to you if you'd like to verify this analysis for yourself.[19]

Figure 4.10 shows mean global surface temperature estimates from the latest generation of models from the IPCC's Phase 5 Coupled Model Intercomparison Project (CMIP5) archive, together with observed global temperatures. This CMIP5 time series summarizes a tremendous human effort – 221 simulations performed by 39 different models or model combinations. While there are differences among the various simulation results, they tend to be fairly small. It is amazing how well the

---

[19] I like to use Geert von Oldenburg's excellent Climate Explorer site, climexp.knmi.nl. There are many other excellent portals that are also available.

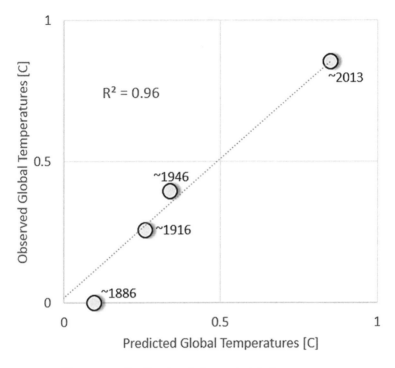

Figure 4.11 Predicted and observed global temperatures, averaged over thirty-year periods.

simulation results track closely with the observed changes in global temperatures, from 1861 to today. Remember, there are no observations used in the climate change simulations, beyond changes in greenhouse gasses, aerosols, and land cover. If the model physics were not basically correct, it is very unlikely that the simulated global temperatures would track so closely with the observations. The CMIP5 time series is a prediction, based on the physics of radiation transfer and the fluid dynamics of the atmosphere and oceans. At a global level, the predictions and observations agree very well. Figure 4.11 shows thirty-year averages of the observed and predicted global temperatures expressed as a scatter plot.

Since graduating from college in 1989, my entire working career has centered around different aspects of data analysis and prediction: econometric modeling in Chicago followed by drought early warning to support famine prevention in Africa. Predictions don't get better than those shown in Figure 4.11. Almost all the variance (96%) of the thirty-year average temperature changes is explained by the model predictions. The slope between the predictions and observations, furthermore, is almost exactly 1. Remember, the models don't use any observed climate information. At a global/decadal scale the models capture *very well* the long-term changes in the Earth's temperature.

The models and observations converge on a story of rapid change, over the course of a few generations. For example, my grandfather's grandfather, Bernard "Barney" Funk volunteered for service in the Union army in 1863 during the Civil War. He served as a mule driver in the Cumberland Army, and fought in the terrible Battle of Chickamauga[20] on September 20, 1863, in which 34,624 soldiers were killed. Returning alive to Indiana, Barney received forty acres of land from the US government and started to farm. Between that bygone era and today, the combined impact of land cover changes and greenhouse gas emissions has warmed the Earth by about 1°C. The last time the Earth was this hot (125,000 years ago, during the last interglacial period), sea levels were about 20–30 feet higher than they are today[21] and hippos[22] lived where Great Britain is today.[23] Measuring this change in generations, that's about 0.2°C of temperature increase per Funk clan member (Bernard begets Edward begets William begets Robert begets Christopher) or 1°C in 150 years. If we run the models forward, they predict that if we don't cut the addition of greenhouse gasses to the atmosphere very soon, then the next 1°C (or more) increase is coming in just the next *thirty* years. The sins of us fathers will be visited upon

---

[20] Chickamauga is an old Cherokee word that means "River of Death."
[21] Hoffman, Jeremy S., et al. "Regional and global sea-surface temperatures during the last interglaciation." *Science* 355.6322 (2017): 276–279.science.sciencemag.org/content/355/6322/276.full.
[22] www.iflscience.com/plants-and-animals/last-time-earth-was-hot-hippos-lived-britain-s-130000-years-ago/.
[23] The temperature maximum during the last interglacial (125,000 years ago) was due to changes in solar radiation and the relative location of the Sun, not greenhouse gas forcing.

the heads of our children and their children. As we discussed in Figure 4.6, terrestrial air temperatures have already climbed 0.6°C between the 1990s and 2015–2019.

> ### Sidebar: The Thomas Fire
>
> Saturday, December 16, 2017, en route from New Orleans to Chicago. Instead of flying to Washington, DC for my planned meetings with the US Agency for International Development and NASA, I have rebooked my flights and am scrambling to get home to my wife, kids, invalid mother, and in-laws: all suddenly potentially in the path of the just reinvigorated Thomas Fire, which suddenly, after days of laying low, erupted early Friday morning into a cacophony of flame. We are not supposed to have forest fires in winter, our wet season. To keep the fire from burning Santa Barbara, firefighters have driven it up into the hills, and due west along the ridge, right toward my house. Crap. In hours, despite a virtual army of firefighters, bulldozers, and fire engines, driven by 70 mile-an-hour winds, very high fuel loads, critically low fuel moistures, and single-digit relative humidities, the Thomas Fire jumped miles to the west. Now directly behind Santa Barbara, it sent fingers of flame and burning coals questing down into the edges of Montecito and the Riviera, right by my in-laws' house.
>
> A few days ago I had made a difficult – and poor – choice, deciding to put business ahead of family, attending a science meeting in New Orleans, thinking that the risk of fire had abated. Boy, was I wrong, and wow was my wife rightly pissed off. Here's the transcript from my cell phone group chat from that morning. Sabina is my wife. Yesi and Matt are a couple, and good friends. She is a volunteer firefighter. Anneli and Nicholas are also good friends. We all live in the mountains behind Santa Barbara, and have kids about the same age. Rochelle is Sabina's mom. Ami is my daughter.
>
> **Around 6:30 AM PST, December 16, 2017**
>
> YESI: 8369 FIREFIGHTERS, 1,102 FIRE ENGINES, 78 DOZERS, 32 HELICOPTERS.
>
> SABINA: AND THE HELICOPTERS! SIKORSKYS, SKELECOPTERS, BIG ONES LITTLE ONES AND ALL COLORS. I WANT TO COLLECT THEM ALL.

ANNILI: Wow

YESI: SAB – as the crow flies, the fire edge is about 3.4 miles away from your Dad's house (☹)

SAB: That's what it looks like! Watching them drop – it's continuous and mesmerizing.

MATT: Reminds me of the Gap fire. We were watching the helos from Camino Rio Verde as the were dropping directly north of our house (☹)

SAB: Opening this channel yes – and including my folks.

ANNELI: How is air

SAB: (pastes picture of horrible cloud of smoke). Air is great now but take a look at this! That will crash down on us soon.

ME: Wow. I have my fingers crossed for everyone.

NICHOLAS: Is the picture taken from the Riviera?

SAB: Pict is from our house.

YESI: Yes. It's not good. I have a mtg with Trout Club crew at 8:30. I'll let you all know. These winds are supposed to be up all day.

ME: OMG.

Now about 8:00 AM PST

Yesi: Jumped San Ysidro Creek now. Montecito winds are 29 miles per hour, gusting to 70.

Nicholas: Stay out of harms way.

Yesi: Mandatory evac for east of highway 154 to Mission Canyon.

Sab: (pictures) Montecito Peak burning.

Matt: Scary

Me: Very

Nicholas: Where are you Sabina?

Me: At her parents house I think.

Sab: I'm fine on way to Trout Club.

Rochellle: Is Sabina safe?
Matt: Yes, Yesi is talking with her now. We're going to head off the mountain.
YESI: We are taking Ami to my father-in-laws in Hope Ranch. We are only volunteer evacuation, but I'd rather they all be safe. I need to be with the volunteer firefighters.
ME: I've rebooked my flight to DC to Los Angeles. Will get into LAX about 9 and Santa Barbara at 11-12.

Now I'm aboard United flight 430. I use the onboard WiFi to download the 12:00 AM Mountain Standard Time MODIS fire imagery.

Me – via e-mail from the plane:

> Looking at latest MODIS Fire imagery - low connectivity, but the fire seems to have reached Casa de Herrera in Montecito, and jumped west of the Montecito fire department. The museum in the center left is the Museum of Natural History.

As freaked out as I was, it was nothing compared to my wife Sabina. Here's the picture she took from the deck of her parents' house that afternoon (Figure 4.12).

A week earlier, on Saturday, December 9, I drove from my house ("a" on the map) up along East Camino Cielo. East Camino Cielo is a beautiful, winding skinny road that rests right on top of the transverse range behind Santa Barbara. It offers stunning views of the Pacific Ocean to the south and the Santa Ynez river valley to the north. After about an hour's drive, I made it past Montecito, to about the place marked "b" on the map below. This figure shows the official evacuation zones on December 6, 10, and 16. On December 8, as I stared eastward down the transverse range, I could only see distant rising smoke clouds to the east and north of Carpentaria. Then, on December 10, a second flare-up occurred, and the fire spread rapidly to the west. I have lived in Santa Barbara for a long time, and the way this fire moved was beyond startling – in aperiodic cataclysmic waves. On December 5 and December 10, the fire expanded by

Figure 4.12 This photo was taken on December 16, 2017, at about 3 PM PST from near the Gibraltar Road and East Camino Cielo intersection. It is looking east toward the community that sits near the top of Gibraltar Road. Image credit: Ryan Cullom/RLC Photography

more than 60,000 acres (each day).[24] Sixty thousand acres is 94 square miles – an area larger than Boston. On December 10, up in the hills, all hell was breaking loose as Santa Anna winds picked up, combining with winds generated by the ferocity of the fire itself.[25] Towering pyrocumulous clouds, or flammagenitus, formed above unstable columns of rising air. Capable of collapsing at any time, these towers could send showers of sparks and whipsaw winds hundreds of yards, posing a serious threat to firefighters. My lovely in-laws, Maurizio and Rochelle Barattucci, live in the foothills at the base of Santa Ynez Mountains ("c" on the map). Saturday night (December 10) was a horrid surreal experience as the fire, driven by 70 mile-an-hour winds, suddenly expanded at an insane rate, eating an acre a second. As I sat on a westbound plane watching satellite fire

[24] upload.wikimedia.org/wikipedia/commons/a/af/2017_12_11–08.57.46.111-CST .jpg.
[25] en.wikipedia.org/wiki/Thomas_Fire.

Figure 4.13 Major advances of the Thomas Fire.

imagery from space, the fire first tore north up into the transverse (east–west) mountain range north of Ventura. Then, in the wee hours, the witching hours, about 3 AM, the wind changed again, blowing from east to west. The fire quickly spread west, first gobbling the chaparral covering the hillsides. The next to fall were the nurseries, avocado orchards, and outlying homes near the city of Carpentaria (Figure 4.13).

On December 11, though, the fire lines held. Forty miles away from Ventura and twenty-five miles from Carpentaria, the ash fell thick on our driveway, covering the cars, the porch, the sidewalks, and the succulents. We wore masks. The crows cackled from the broken oak branches. Their cries echoed across a gray mezzotint landscape drawn by Edgar Allen Poe. Then again, on December 16, the fire expanded rapidly again. Amazingly, a literal army of firefighters managed to protect Montecito and Santa Barbara.

5 TEMPERATURE EXTREMES – IMPACTS AND ATTRIBUTION
Shocks, Exposure, and Vulnerability

Our planet is warming rapidly. Global warming is already hurting people, contributing to widespread increases in health risks associated with extreme temperatures.[1] The impacts of climate change are arising through the potentially explosive interaction of **weather shocks, exposure,** and **vulnerability** (Figure 5.1). The shock-exposure-vulnerability construct is a common framework for understanding the impact of extremes. Shocks represent weather-related extreme events. Exposure quantifies the potential losses associated with these events. Population, economic investment, and ecosystem productivity are commonly associated with increased exposure. Vulnerability represents a lack of capacity to resist negative impacts. Almost everywhere we see very rapid increases in impacts due to multiplicative combinations of expanding shocks and exposure. More frequent heat waves and precipitation extremes impact rapidly expanding populations. More expansive fires threaten growing rural and semi-rural settlements. Cyclones and tidal surges inundate

---

[1] Watts, Nick, et al. "The 2019 report of The Lancet Countdown on health and climate change: ensuring that the health of a child born today is not defined by a changing climate." *The Lancet* 394.10211 (2019): 1836–1878. www.thelancet.com/journals/lancet/article/PIIS0140-6736(19)32596-6/fulltext.

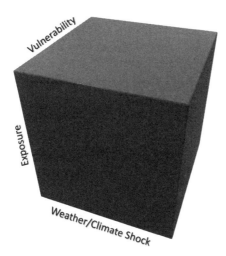

Figure 5.1 Dimensions of hazards or risks associated with weather and climate extremes. Impacts arise through the interaction of three distinct factors: the magnitude and frequency of climate and weather shocks, the underlying level of exposure in the human and natural systems experiencing those shocks, and the underlying vulnerability of these systems.

increasingly populated, valuable, and vulnerable coastline properties. And in each of these cases, the poor and vulnerable tend to bear the brunt. Adopting a card game analogy, weather deals the cards, exposure measures how much you have to lose, and vulnerability quantifies how well you can absorb that loss.

Of all weather hazards, heat waves tend to be the most immediate, and potentially the most deadly. All living organisms must maintain subtle balances with their surrounding ecosystem. Extreme temperatures, especially under humid conditions, can lead to heat stress and heatstroke, acute kidney injury, and dramatically increase the chance of dying due to heart problems, especially for older people.[2] Heatstrokes occur when rising internal body temperatures impact your nervous system. When the air surrounding your body reaches extreme levels, you may

---

[2] Székely M., Carletto L., and Garami A. "The pathophysiology of heat exposure." *Temperature (Austin)* 2 (2015): 452.

not be able to sweat fast enough to cool down. In response, seeking to maintain a balanced temperature, your body moves more blood toward your skin, away from your internal organs. This can damage these organs while also putting strain on your heart, potentially leading to heart problems. The brain receives less oxygen. More toxins enter your bloodstream and kidneys.

Cities, which are often warmer than the surrounding countryside, can expose millions of people to heat waves. The elderly and children are generally much more vulnerable. In particular, during periods of extreme heat, young children have a greater risk of electrolyte imbalance, fever, respiratory disease, and kidney disease.[3] Extreme temperatures can also increase the chance of interpersonal violence,[4] collective civil conflict,[5] and dramatically decrease worker productivity, which can negatively impact national-level productivity. Extreme temperatures can stunt childhood growth and stress pregnant mothers and their babies, contributing to increased frequencies of low-birthweight babies that often have serious health problems. In crop-growing regions, extreme temperatures can increase water demand, alter growth rates, and directly damage plants, especially during the critical germination and grain-filling periods in the middle of a growing season. In the ocean, marine heat waves pose severe threats to coral reefs, fisheries, and the human and nonhuman populations that depend on them.

Many warm regions of the globe are densely populated, resulting in high levels of exposure to heat extremes. Historically, many of the most ancient civilizations centered around hot, low subtropical river basins: the Nile, Tigris-Euphrates, Indus, and Ganges River systems helped give birth to humanity by providing stable sources of water. Terrestrial plants and oceanic phytoplankton are light-absorbing life-making miracle machines,

[3] Xu Z., Sheffield P. E., Su H., Wang X., Bi Y., and Tong S. "The impact of heat waves on children's health: a systematic review." *International Journal of Biometeorology.* 58 (2014): 239–247.

[4] Sanz-Barbero B., Linares C., Vives-Cases C., González J. L., López-Ossorio J. J., and Díaz J. "Heat wave and the risk of intimate partner violence." *Science of the Total Environment.* 644 (2018): 413–419.

[5] Levy B. S., Sidel V. W., and Patz J. A. "Climate change and collective violence." *Annual Review of Public Health.* 38 (2017): 241–257.

constantly transforming carbon dioxide, water, and incoming solar radiation into sugars and carbohydrates. Given abundant water and carbon dioxide, this transformation process tends to be the most productive where and when we receive the most incoming solar radiation. This is also where and when temperatures tend to be the warmest. So, on land and in the oceans, many of the most biologically productive tropical and subtropical regions face high levels of risk from rising temperatures. These risks may impact the organisms absorbing the most carbon dioxide from the atmosphere, creating potential "positive" climate change feedbacks that can help drive the bus even faster over the cliff. Many of the warmest regions, like topical forests or coral reefs, team with life that helps sequester carbon dioxide. Yet these very warm regions can also be dramatically damaged by modest increases in temperatures. Focusing on human impacts, we will see that billions of humans face annual threats associated with the north–south passage of the summer monsoons and the areas of maximum of incoming solar radiation.

Small Changes Can Dramatically Increase the Chance of Extremes

Before going forward, we need to discuss further the interaction between weather and climate extremes and hazards. Weather and climate extremes are either short-term (weather) or long-term (climate) anomalies that are "extreme" based on the historical behavior of some phenomena. For example, in early November 2019, the New South Wales region of Australia and the city of Sydney faced catastrophic fire danger fueled by very poor rains over the last twelve months, exceedingly rapid offshore winds blowing from the interior of the continent toward the Tasmanian Sea, and exceptionally warm air temperatures (see sidebar). High winds and warm temperatures qualify as extreme events, and their combination can be exceptionally dangerous. The potential impact of these *weather* conditions can be exacerbated by lower-frequency *climate* conditions, such as a year's worth of very warm air temperature and very low rainfall.

Climate scientists often use "percentiles" to describe weather and climate extremes. At a given location and time, a weather variable, such as daily maximum air temperatures, will have a distribution of possible values. A bell curve–like normal distribution is one common distribution that many people are familiar with. The "percentile" values within this distribution range from the lowest possible (the 0th percentile value) to the greatest possible (the 100th percentile value). A percentile value of 50 would correspond to the median (i.e., middle) value; half the observed values would be expected to be above and below this value. When dealing with extremes, we often deal with values like the 99th percentile maximum air temperature. If we measure each day's warmest air temperature for a year, and sort the 365 warmest days, then the average of the third and fourth warmest days would correspond to the 99th percentile value for that distribution. When discussing climate change, we often examine changes in the values associated with a given percentile. Alternately, we can pick a threshold (say, a temperature threshold of 40.6°C/105.1°F) and look at changes in the frequency of values exceeding this threshold.

To understand the influence of climate change on extremes requires an understanding of both the Earth's seasonal cycle and how small changes in average conditions can translate into big changes in weather extremes. When you watch or read about this month's extreme temperatures, think about how the locations of these extremes align with the passing seasons. Every year the latitude of maximum incoming solar radiation moves north and south with the tilting of our planet in its annual rotation about the Sun. Areas near the equator receive the most radiation, but also tend to be cloudy. These clouds reflect the sunlight back into space. So the warmest maximum daily air temperatures tend to be located in the subtropics, between about 17° and 37° north and south, in the summer hemisphere. During summer in the northern hemisphere, we find that the warmest (99th percentile or greater) air temperatures have very warm average values of ~44°C (111.2°F). As the point of maximum incoming solar radiation dips south of the equator, a second band of maximum heating arises in the southern hemisphere.

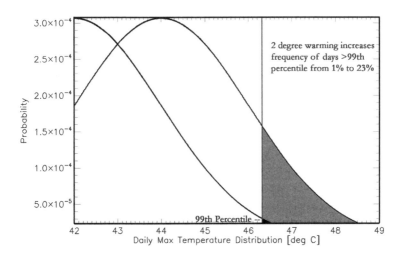

Figure 5.2 Normal distribution curves for a hot location on Earth.

Putting these concepts together, we can imagine that a typical very warm region in the northern or southern subtropics might have a mean value of ~42°C for the warmest (99th percentile) days in a year. Typical values ranging about this mean may go from about ~39.5°C to 46.5°C (103.1 –115.7°F). The warmer half of this distribution is shown in Figure 5.2. The most probable value (42°C) is shown on the left, and then the probability curve drops rapidly. The shaded (black) region on this curve corresponds to the 1 percent of distribution corresponding to values exceeding the 99th threshold (46.3°C). Now we add 2°C and redraw the corresponding bell curve. The chance of exceeding the 99th percentile value of 46.3°C jumps 23-fold. *A relatively small change in the mean can result in a very large increase in the number of extreme events.*

How Extreme Have Temperatures Been in 2015–2019?

The amount of daily station data in many parts of the world is quite limited. But we can use the NASA Goddard Space Flight Center's 1880–2019 monthly temperature data set to

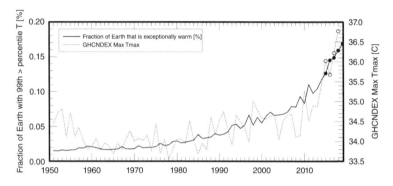

Figure 5.3 Time series of the fraction of the Earth with exceptionally warm temperatures (solid black line) and globally averaged maximum warmest-day temperatures (dashed line). Circles identify the 2015–2019 period.

explore the changes in temperature extremes.[6] This data set has sea surface temperature estimates over the ocean and two-meter air temperature estimates over land (both measured in degrees Celsius). This data set is routinely used to assess global temperatures. The 2015–2019 annual global land-sea temperature anomalies were each warmer than all the previously observed values. To track areas with exceptionally warm conditions, for each month and for each year we can identify areas with ocean or air temperatures in the top 1 percent of the 1880–2019 record. Then for each year, we can translate this information into a fraction between 0 and 1, corresponding to the fraction of the Earth with temperatures in the top 1 percent of the historical record. The black line in Figure 5.3 displays this data. Black circles denote values for our years of interest.

This time series should deeply disturb you. The fraction of the Earth with exceptionally warm ocean or air temperatures is increasing very rapidly. When I was born, it was about 2 percent. By the time I started graduate school, it was around 5 percent. When my children were born, it hovered around

---

[6] data.giss.nasa.gov/gistemp/.

7 percent. Since that time the area of exceptional warmth has doubled again to around 17 percent, increasing rapidly just between the beginning and end of our focus period (2015–2019). Now almost *one-fifth* of the globe experiences extreme temperatures at any given time. Extreme temperatures are rapidly becoming much more frequent. According to Maximiliano Herrera, a Costa Rican climatologist dedicated to rigorously tracking global temperature extremes using station data,[7] the 2015–2019 period has brought record-breaking temperatures at 384, 337, 201, 430, and 619 weather station locations, year to year.

These temperature extremes are having severe health impacts. Between 2015 and 2019, the Emergency Events Database (EM-DAT),[8] maintained by the Centre for Research on the Epidemiology of Disasters, identified 71 extreme-temperature disasters, related to 9,116 deaths, 90,014 injuries, affecting 4.5 million people and resulting in $1.8 billion (€1.64 billion) in damages. According to EM-DAT, in 2015 temperature extremes in France, India, and Pakistan claimed 3,275, 2,248, and 1,229 lives, respectively. In 2016, 2017, and 2018, extreme heat waves in Morocco affected 750,000, 1,650,000, and 70,000 people, respectively. In 2018, extreme heat waves injured 49,000 people in Japan. In 2016, a heat wave in China resulted in $1.6 *billion* dollars in losses. In 2019, 173 people died in Japan and 112 people died in India due to extreme temperature–related causes.

As we now routinely experience, the impact of this warming often follows the maxima of solar insolation. During the northern hemisphere summer of 2018, extreme temperatures meant that drought and wildfires threatened Europe. In January 2019, extreme heat stretched across the southern hemisphere, and forty-four fires, fueled by extremely hot and dry conditions, burned for weeks across Tasmania.[9] In June and July 2019, Europe experienced record-breaking heat waves, and extremes

---

[7] www.mherrera.org/temp.htm.   [8] www.emdat.be.
[9] earthobservatory.nasa.gov/images/144486/fortnight-fires-in-tasmania.

were linked to some 1,800 deaths in France.[10] In May and June 2019,[11] a severe heat wave struck India and Pakistan. Temperatures in northern India, in Rajasthan, reached 50.8°C (123.4°F). Consulting India disaster statistics in EM-DAT, the international emergency database, reveals this "extreme" to be part of a pattern of the new normal. According to EM-DAT, 112 deaths were associated with the 2019 heat wave. Heat waves were linked to 264 deaths in 2017 and 300 deaths in 2016. And in 2015, a terrible heat wave was tied to 2,248 deaths. As the locus of maximum solar radiation shifted south, Australia faced, yet-again, record-breaking temperatures and widespread wildfires (sidebar).

Using daily weather station data, we can also calculate the maximum daily temperatures at each weather station location, and then average these values over the globe. Several groups of researchers are committed to the important task of monitoring these types of climate extremes, and we can access many of their valuable datasets online at www.climdex.org/.

The dotted line in Figure 5.3 shows estimates of global averaged Global Historical Climate Network Index (GHCNDEX)[12] annual maximum temperatures. This dotted line may be even more concerning than the solid black line in Figure 5.3. Between the 2000s and our 2015–2019 period of interest, the maximum terrestrial air temperature values appear to have increased by something like +1.5°C. Daily temperature maximums tend to occur in the afternoon, as the daily heating from the Sun accumulates. Seasonally, these maxima tend to occur in the late spring or summer, when again, the heating by the Sun tends to reach maximum values. In just a few pages, we will provide model-based evidence suggesting that this trend will continue. Such increases in extreme temperatures will be very dangerous for people, crops, forests and livestock. Such

---

[10] www.bbc.com/news/world-europe-49628275.
[11] www.thelancet.com/journals/lancet/article/PIIS0140-6736(19)32596-6/fulltext.
[12] Donat, M. G., L. V. Alexander, H. Yang, I. Durre, R. Vose, and J. Caesar. "Global land-based datasets for monitoring climatic extremes." *Bulletin of the American Meteorological Society*. 94 (2013): 997–1006.

warming will shift the distributions of extreme events to the right, as illustrated in Figure 5.2.

Quantifying Increases in Exposure to Extreme Air Temperatures

How many people may be exposed to exceptionally warm temperature extremes? This important question can be difficult to address with rigor, especially because accessible weather station data tends to be very sparse in many countries. To help overcome this limitation, my research group has developed a new maximum temperature product[13] that uses a long time series of geostationary weather satellite observations[14] and Berkeley Earth[15] weather station data to provide accurate estimates of maximum air temperatures. These estimates are particularly well suited to monitoring conditions in places like Africa and India, regions where vulnerability is high, exposed populations are large and rapidly increasing, and typical extreme temperatures are already very high. As context, we can examine a map (Figure 5.4) showing the mean daily 99th percentile air temperatures, based on a new dataset I have created with my team at the University of California Santa Barbara. This new high resolution (0.05 degree ~ 5 km²) data set is referred to as the Climate Hazards Center Infrared Temperature with Stations T_{max} (CHIRTS$_{max}$) dataset. At present it extends from 1983 to 2016 and is in the process of being updated. It benefits from lots of station data augmented by long continuous set of satellite-based observations of the Earth's surface emission temperatures. This makes it particularly well-

---

[13] chc.ucsb.edu/data/chirtsmonthly, Funk, Chris, et al. "A high-resolution 1983–2016 T max climate data record based on infrared temperatures and stations by the Climate Hazard Center." *Journal of Climate* 32.17 (2019): 5639–5658. https://www.chc.ucsb.edu/data/chirtsdaily. Verdin, Andrew, et al. "Development and validation of the CHIRTS-daily quasi-global high-resolution daily temperature data set", Scientific Data, 7.30 (2020). https://www.nature.com/articles/s41597-020-00643-7.
[14] www.ncdc.noaa.gov/gridsat/   [15] berkeleyearth.org/data/

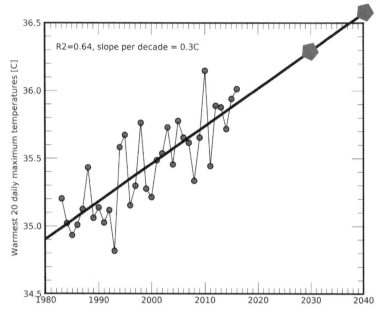

Figure 5.4 Population-weighted averages of the warmest 20 days of each year. Based on the CHIRTSmax dataset and 2020 Gridded Population of the World population estimates.

suited to look at changes in data-sparse areas – those areas where people also tend to be highly vulnerable.

Broadly following the approach taken in a recent Lancet article on the increasing extreme heat risks, we can use the daily $CHIRTS_{max}$ data set and gridded population estimates to explore the effects of changes in air temperatures during each year's warmest days (Figure 5.4). Before even asking what this time series means, linger for a moment considering what it involves. The population data involves an incredible effort involving hundreds of national censuses, remote sensing, and sophisticated integration and modeling. Swap in weather station observations for census estimates, and the same statement holds for air temperature estimates. These datasets, compressing hundreds of lives worth of human effort into collected sets of information, provide the basis for 'science haikus' like Figure 5.4. More such figures will follow.

Figure 5.4 tells us it is getting a lot hotter quickly in places where people are living, during the hottest ~5 percent of the year (~20 days). A simple trend line associated with this data explains most (64 percent) of the year-to-year variability. This trend (about 0.3 °C per decade), works out to a ~+1.2 °C increase between 1980 and 2020. That is a lot of warming. Extending this trend to 2030 or 2040, we would find additional temperature increases of +1.5 and 1.8°C. These outcomes are marked with gray polygons.

The current mean of the warm temperatures is around 36°C, but there are many places that are much warmer. And many of these places are heavily populated, tropical or subtropical, and at relatively low elevations – the perfect recipe for high levels of exposure combined with frequent extreme heat waves Many of these regions are historic river-based cradles of civilization. The Niger River flows east across western Africa, through Guinea, Mali, Niger and Nigeria. The Nile flows north from Ethiopia and Uganda through South Sudan, Sudan and Egypt. The Tigris-Euphrates river system that begins in the Zagros Mountains of eastern Turkey flows through Syria and Iraq on its way to the Persian Gulf. The Indus River flows south from the Hindukush mountain range through Pakistan, while the Ganges River system flows east across northern India to drain into the Bay of Bengal. For thousands and thousands of years, generation after generation, humans have lived in these hot valleys, giving birth to civilization along the way. In 2019, 2.3 billion people lived in these countries, almost one out of every three humans. By 2050, United Nation's population projections anticipate an additional billion people in these regions, a 50 percent increase. Those 3.3 billion people will be exposed to much warmer extreme heat waves.

To help visualize the billions of people in harm's way, we can examine a simple map of daily 2000–2016 99th percentile maximum temperature values (Figure 5.5). In many parts of India and Pakistan, temperatures reach incredible levels, exceeding 44°C (111°F), and reaching as high as 47°C (117°F). These very warm regions tend to align with river basins, like the

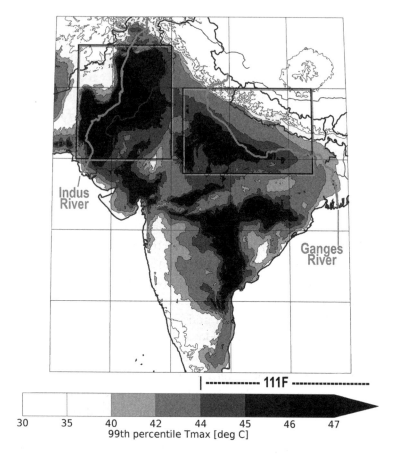

Figure 5.5 Map of 99th percentile 1983–2016 air temperatures for 2000–2016 and 2050 (right), based on a 'business as usual' climate change change scenario based on the 8.5 Wm^{-2} Representative Concentration Pathway scenario

Indus and Ganges watersheds, which are the most densely populated locations on the planet. In between these watersheds we find higher elevations and cooler temperatures. The United Nation projects that India and Pakistan's 2050 population will be two billion. Boxes around the Indus Basin in Pakistan and the Indo-Gangetic plain in northern India highlight these high risk regions. Climate change projections (discussed below) indicate imminent increases in mean temperatures that will place these extremely large populations in great peril.

We can quantify the risks associated with heat waves by using the high resolution daily $CHIRTS_{max}$ data to identify heat waves, and then counting the number of people exposed to these extremes. This provides a numerical measure of the interaction of extreme weather and exposure, two of the main dimensions of climate hazards (Figure 5.1). There are many approaches to identifying heat waves. Most definitions require temperatures to exceed a given threshold for several days (often three or more). Persistent warm conditions are much more harmful biologically. Thresholds may be based on daily minimum or daily maximum temperatures. Daily minimum temperatures occur at night, and very high nighttime temperatures reduce the body's ability to recover from heat stress. It is also common to use measures of "apparent" temperature that take into account the effect of atmospheric humidity. Higher humidity levels make it harder for humans to cool their bodies through perspiration. Here we will use a very simple definition of a heat wave as a three day period in which the temperature for each day is warmer than 40.6°C (105.1°F). Using this threshold we can calculate the number of heat waves at each location in each year between 1983 and 2016. Using current gridded population estimates, we can calculate the population weighted average of these counts. This time series of global heat wave counts (not shown) exhibits a strong upward trend of ~0.5 events per decade, increasing by about 40 percent between the early 1980s and late 2010s, from four to about five-and-a-half events per year.

To estimate exposure to heat wave risks, we can combine our heat wave counts with gridded annual population estimates for every year between 2000 and 2016 (Figure 5.6). These results are very concerning, and indicate very large increases in the number of people exposed to heat extremes. Our analysis begins in 2000 because gridded population estimates prior to that year are less reliable. For each year, we multiply the number of heat wave events at a given location by the population in that location. The result of this multiplication produces values in units of "people-events," capturing the interaction between population growth and increases in

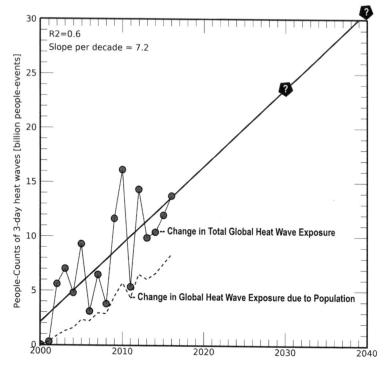

Figure 5.6 Time series of annual global estimates of total heat wave exposure (gray circles). The dashed line shows just the estimated exposure values associated with variations in population.

air temperatures. Then for each year we add up the total people-events for the planet. The time series marked with grey circles in Figure 5.6 indicates very rapid increases in people-events. A trend of ~7.2 *billion* people-events per decade means that a relatively modest shift in extreme temperatures (Figure 5.4) is substantially increasing the frequency of very warm events (depicted schematically in Figure 5.2), and interacting with population growth to create billions more exposure events every decade.

By fixing the population at 2000 levels we can estimate the temperature component and remove it, isolating the population-driven component of the increases in exposure people-events. The population component is identified by the dashed line in Figure 5.6. If the total number of people-events increased by around 14 billion since 2020, population growth alone accounted for about 8 billion while increases in maximum air temperatures accounted for about 6 billion. The interaction between atmospheric warming and population increases is already creating dramatic expansions in human exposure. These trends will absolutely continue, for at least the near-term future.

Both the climate and human socioeconomic systems exhibit strong levels of persistence. Over the next several decades we are certain to see continued warming and population growth. Greenhouse gasses build up in the atmosphere and persist for many years, while the ocean continues to store more and more thermal energy. At present there are almost a billion young women on the planet,[16] and most of these women will become mothers, so demographers anticipate rapid population growth. And progress toward reducing greenhouse gas emissions remains very slow. If the observed trend in heat exposure continues, we might see a 2000–2030 increase in exposure of 24 billion people-events, and 2040 increases exceeding 30 billion people-events a year. Many more people will experience many more extreme heat waves over the next two decades. These heat waves are already having serious impacts on health, especially for children, pregnant mothers, and the elderly. Dangerous climate change is not a "what if" but a "happening now" with "a lot more coming soon."

Can We Attribute a Portion of These Impacts to Climate Change?

Attribution science involves analyzing the root causes of phenomena, such as human diseases or extreme weather patterns. Such analysis can lead to prediction and prevention. For

---

[16] social.un.org/youthyear/docs/fact-sheet-girl-youngwomen.pdf.

example, in 1854, the British physician John Snow correctly attributed the outbreak of cholera in London to a contaminated pump at the intersection of Broad and Cambridge streets. Removing the handle to this pump stopped the epidemic. Motivated by a similar desire to predict and protect, meteorologists have spent decades exploring the mechanisms that cause weather and climate extremes. Unlike medical researchers, who can often use structured medical trials complete with placebos and control groups, Earth scientists face severe experimental limitations. We only have one world, and can't run experiments on hurricanes and heat waves. Extreme events, furthermore, are by definition rare, and typically arise through complex non-linear interactions. Yet as described in a recent report by the United States National Academy of Sciences[17]: "Effective, rigorous, and scientifically defensible analysis of the attribution of extreme weather events to changes in the climate system not only helps satisfy the public's desire to know but also can provide valuable information about the future risks of such events to emergency managers, regional planners, and policy makers at all levels of government." The new science of extreme weather and climate event attribution seeks to provide such analysis.

Framing attribution questions correctly is a critical first step; we need to avoid the question "did climate change cause X," where X is an extreme event. Climate change alters the amount of heat in the atmosphere, land, and oceans, the distribution of greenhouse gasses and aerosols in our atmosphere, sea levels, sea ice, and ocean acidity. These changes can modify the frequency or intensity of weather and climate change events, but are not typically the primary cause of anything. To return to our analogy of playing cards, climate change might shift the distribution of cards, making some hands more likely than others.

---

[17] www.nap.edu/catalog/21852/attribution-of-extreme-weather-events-in-the-context-of-climate-change.

Understanding the influence of such distribution changes on temperature extremes is relatively straightforward. Of the range of weather and climate extremes (e.g., droughts, floods, fires, hailstorms, tornadoes, hurricanes, etc.), the influences of climate change on extreme heat events are the best understood, and we have the highest level of confidence when examining these types of events. This confidence arises from both the relative simplicity of extreme heat events as well as our ability to model these influences. For example, we are all familiar with the idea that weather systems are produced by the movement of high- and low-pressure systems. High-pressure cells typically produce subsidence (sinking motions) and cloudless skies that allow more sunshine to reach the Earth. These and other naturally occurring factors are represented well by our current generation of climate models, and our models do a good job of predicting day-to-day and decade-to-decade variations in air temperature. We can contrast this predictability with wildfires and cyclones, which arise through very complicated small-scale and highly nonlinear interactions. Both our theoretical understanding and our ability to model and predict these highly complex systems is limited. Hence, it is much easier to understand how one degree of warming might change the distribution of extremely warm days than how one degree of warming might change the number, extent, or intensity of wildfires or hurricanes.

By combining our observed high-resolution temperature and population data with climate model–based estimates of temperature changes, we can explore two important questions: (1) What would temperature exposure levels look like in a world without climate change? (2) What will likely future temperature exposure levels look like in a world with climate change? We address both of these questions using a climate analysis technique called "perturbation." We start with a huge amount of data that describes the world as we know it over the 2000–2016 time period, the data summarized in Figure 5.6. Then we perturb the observed temperatures by model-based temperature deviations, based on large collections (ensembles) of climate change simulations.[18]

Climate models describe the physics controlling the motions of the atmosphere, ocean, and land. These model components influence each other. Today's wind patterns and cloud distributions will subtly affect tomorrow's ocean temperatures. These ocean temperatures will then influence future weather patterns. And so on. Day in and day out, year after year, these models can be used to simulate our weather and climate with a pretty exciting degree of realism, thanks to decades of careful improvements made by a global collection of modeling centers. Massive efforts in support of initiatives like the Intergovernmental Panel on Climate Change (IPCC) have generated petabytes of simulations that can be used to assess the influence of climate change.

One great advantage of model simulations is that scientists can alter the components that control the Earth's "radiative forcing." Radiative forcing refers to those aspects of Earth–Sun system that control our basic radiative balance. There are important natural factors such as the magnitude of incoming solar radiation (modulated by Sun spots on short time scales) and the cooling effect of volcanic eruptions. There are also important anthropogenic (human-caused) variations such as the distribution of greenhouse gasses and aerosols, and alterations to the Earth's land surface. Greenhouse gasses and aerosols change how energy moves through the atmosphere. Alterations to the Earth's surface can alter how much solar radiation gets reflected back into space. Here we explore three sets of climate change simulations representing different global warming. One set, based on the 4.5 W/m^2 Representative Concentration Pathway (RCP4.5), represents a future in which we effectively reduce our carbon emissions. A second set, RCP8.5, represents a future in which we continue

---

[18] These results are based on a large set of simulations from thirty-eight to forty-one different models obtained from the KNMI climate explorer: climexp.knmi.nl/start.cgi. Temperature changes were based on the modeled changes in the warmest month of the year.

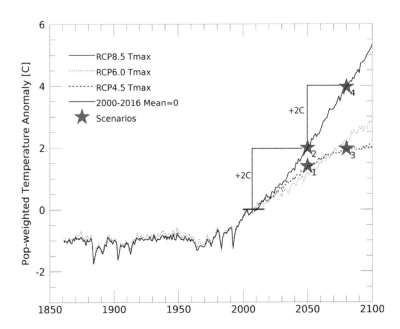

Figure 5.7 Time-series of population-weighted averages of changes in modeled daily maximum temperatures. The time series are expressed as changes from the 2000–2016 time period.

to emit large amounts of greenhouse gasses. A third set, RCP6.0 represents an intermediate warming scenario. The RCP scenarios (described in Chapter 13) correspond to different projections of future emissions.

Figure 5.7 presents the population-weighted average change in daily maximum temperatures, during the warmest month of the year, for RCP4.5, RCP6.0, and RCP8.5 scenarios. These results are presented as changes from a 2000–2016 baseline period, chosen to correspond to our heat wave analysis (from Figure 5.6). In addition to future warming projections, the models have also been forced with historical radiative forcing conditions, so the time series begin in 1861. For each year, we have multiple estimates from approximately forty models (it varies a little by RCP). Figure 5.7 shows the average of these simulations.

There are four important things to be learned from this figure. First, the simulations suggest that we have already experienced about 1°C of human-induced warming. Second, no matter what, we are very likely to experience about another degree of warming by 2050. Even if we adopted vigorous reductions in emissions now, the persistent effect of our accumulated greenhouse gasses will continue to warm the Earth for the next thirty years. But (third very important point) even a modest effort to mitigate our emissions (corresponding to the RCP6.0 scenario) will reduce the amount of 2050 warming by about half, limiting it to around 1°C. On the other hand, continuing on our current rapid (RPC8.5-like) emissions trajectory will see our children mature in a world likely to be two degrees warmer than present conditions. Which leads us to point four. The RCP8.5 scenario is estimated to warm the world inhabited by our childrens' children by 4°C. Committing ourselves to mitigation now can head of this catastrophic increase.

What might life have been like in a world without climate change? We have the tools to answer this question quite precisely, at least for relatively simple phenomena like temperature extremes. We can estimate the spatial pattern of human-induced warming, remove it from the observational record, and estimate the number of people exposed to extreme heat events. This allows us to *attribute* a proportion of the observed extreme heat waves to climate change. We can perform this attribution by perturbing our 2000–2016 temperature data, adjusting it to align with a world without human-induced warming, and then recalculating our risk metrics. In this "counterfactual" experiment we construct a world that might have been by taking the observed 2000–2016 temperature observations and cooling them by the difference, at each location, between the 1861–1900 and 2000–2016 temperatures simulated by our climate change models. This gives us a reasonable approximation of a world like today but without the influence of human-induced warming. We can then recalculate the frequency of extreme heat waves and the billions of people-events experienced, on average, between 2000 and 2016. The bottom left star in Figure 5.8 shows these results.

The answer we find implies that climate change has nearly doubled the number of extreme heat waves. Between

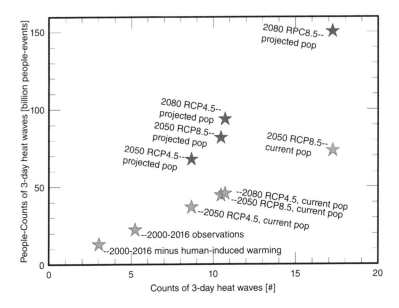

Figure 5.8 Estimates of annual heat frequencies and the number of heat wave people-events (#people x #heatwaves).

2000 and 2016, we observed about five three-day heat waves per person. Subtracting an estimated human-induced warming (about 1°C), we found that we would expect to see about three extreme heat waves per person per year. So, climate change has increased the number of extreme heat waves by about 70 percent. This means that climate change is negatively impacting hundreds of millions of people *now*, enhancing the frequency of severe health risks, increasing conflict, feeding climate migration, and reducing worker productivity. *Now*.

But "climate change" is just shorthand for "our collective actions." We all are implicitly hurting, increasing conflict, feeding crises, and reducing productivity. This may be a fourth moral dimension of climate hazards. In addition to shocks, exposure, and vulnerability (Figure 5.1), we may need to include a dimension associated with climate change attribution – moral culpability. In Chapter 4 we discussed how the incredibly rapid rate of greenhouse gas increases (Figure 2.3) was produced an incredibly rapid rate of global warming (Figures 4.6 and 4.7). In the warmest tropical oceans, marine heat waves are killing coral reefs. In the warmest locations on land, atmospheric heat waves are already killing thousands of people. There are billions of people in harm's way, and human biology is not going to be able to evolve rapidly enough to keep the most vulnerable of these populations from massive increases in exposure.

Explosive Interactions with Future Population Growth

We can next use our projections to explore how changes in exposure might increase in the future. The "2050 RCP4.5, current population" projection, shown with a light gray star, was calculated by warming our observed 2000–2016 climate by the average modeled RCP4.5 increase. This would correspond to the Scenario 1 in Figure 5.7 if our 2050 population remained the same as today. The lower gray stars to the right of this value would correspond to Scenarios 2–4. The RCP4.5 2080 world (Scenario 3) has about the same amount of warming as RCP8.5 2050 (Scenario 2).

These results tell us that if we ignore the impacts of population growth, and just examine the impacts of climate change, we will still see massive increases in the amount of heat exposure. Even the relatively modest 2050 RCP4.5 scenario indicates that the frequency of extreme heat waves will almost double by 2050. With unchecked emissions growth, we could see an almost fourfold increase in the frequency of very extreme heat waves by 2080.

Factoring in both climate change and a moderate population growth scenario,[19] we find huge potential increases in exposure. These results are marked with dark stars. The most modest outlook 'RCP4.5 2050, projected population' indicates

---

[19] These population change estimates were based on gridded population projections corresponding to Shared Socioeconomic Pathway 2. Jones, B., and B. C. O'Neill. 2017. *Global Population Projection Grids Based on Shared Socioeconomic Pathways (SSPs), 2010-2100*. Palisades, NY: NASA Socioeconomic Data and Applications Center (SEDAC). doi.org/10.7927/H4RF5SoP.

a massive increase in heatwave exposure, with the number of events increasing from an observed 2000–2016 baseline of about 20 billion to a 2050 baseline of about 70 billion people-events a year. Even under a scenario of modest population and emissions increase, we are likely to see very large increases in the number of exposure events. By 2080, this same scenario identifies another ~20 billion people-event increase, a magnitude of exposure consistent with the total level experienced today. An RCP8.5 scenario would be catastrophic. The number of heat wave exposure events might increase by 670 percent.

Conclusion

Looking beyond the numbers, we need to extend our imagination, employ our empathy. Billions are currently exposed to extreme heat, in burgeoning cities, often without recourse to air conditioning. As our planet warms, how many will fear the Sun's passage, knowing that it may bring remorseless heat to our hottest cities for days on end? Have you walked the streets of Cairo, Khartoum, Niamey, Bangalore, Bhadgad, or Hyderabad on a very hot summer day? How about Chicago's South Side, or one of the poor banlieues ringing Paris? Block after block, the stands, markets, and apartments spread, each a microcosm of human life. But increase the frequency of extreme heat events, and these burgeoning cities may become extremely hazardous, especially for the poor and elderly who can't afford air conditioning.

The results presented here are deeply concerning. They suggest that the best we can realistically hope for is a doubling in the number of extreme heat waves by 2050. If we adopt rapid reductions consistent with RCP4.5, we may see about nine extreme heat waves per person per year, compared to the five events in recent years. This is three times the frequency in a preindustrial world without climate change.

Fast forward yet another thirty years, and the seeds of our decisions today will truly bear fruit, benign or no. If we adopt a moderate mitigation strategy now, our RCP4.5

estimates indicate that by 2080, conditions will only get a little worse, with nine heat waves per person in 2050 increasing to eleven in 2080. But if we fail to mitigate, and follow the RCP8.5 pathway, we find a stunning increase to seventeen heat waves per person per year, about three times the current frequency and about six times the number in a world without climate change.

But this only begins to tell the story, because these extremes are often happening in areas with rapidly growing populations. When we factor in moderate population projections, we find explosive increases in the number of estimated annual people-events (y-axis, Figure 5.8). By 2050, the optimistic RCP4.5 scenario appears associated with a 300 percent increase in exposure. The pessimistic RCP8.5 scenario indicates almost a fourfold increase in people-events compared to 2000–2016, to a total of 81 billion people-events per year. Going forward to 2080, population growth slows but still continues to increase. Even the modest RCP4.5 scenario calls for some 93 billion exposure events per person per year, and for RCP8.5 our estimate is *150 billion.* Currently, the overall level of exposure (22 billion) is about three times the population of the planet. By 2080, this level of exposure may be ten to fourteen times greater. These are our children we are talking about.

Our bodies, especially the bodies of young children, pregnant mothers, in utero babies, and old people, experience severe stress when exposed to consecutive days of very warm weather. These extremes also feed conflict and substantially reduce worker productivity. Here, in this chapter, we have used some of the best available data to estimate the frequency of extremely warm days, and the overall level of exposure, measured in people-events. These observations are deeply concerning. Projecting into the future, we find even greater reason for concern.

Sidebar – Examining Extreme Temperatures in India and Pakistan

Each year, the Bulletin of the American Meteorological Society publishes a special issue presenting assessments of how human-caused climate change may have affected the strength and likelihood of individual extreme events.[20] While a detailed assessment of these special issues is beyond the scope of this book, describing a single study can help elucidate how expert climate attribution scientists do their work. To assess the influence of human-induced climate change on extreme events, scientists borrow techniques from epidemiology. In the broadest strokes, we can divide epidemiological analyses into situations where we can perform controlled experiments and those where we cannot. For example, if we are studying the efficacy of a pain medication, we might perform a controlled experiment using a large sample of patients. Some patients will take the real medicine. Some will take a placebo. Sometimes, however, we are morally prevented from such experimentation. We would not want to perform such a study on children facing a severe, life-threatening disease. In such a setting we typically have to rely on using statistics. In climate attribution, the first type of experiment is usually performed using climate models. The second type of experiment is performed using historical data.

The India 2015 heat wave attribution study by Drs. Michael Wehner, Daithi Stone, Hari Krishnan, Krishna AchutaRao and Federico Castillo[21] provides a good example of both types of analyses. In May and June 2015, severe heat waves struck New Delhi, Allahabad, Jharsuguda, Hyderabad, and Khammam.[22] In New Delhi, a city with 30 million people, extreme temperatures reached 45.5°C (113.9°F) and the city streets literally melted.[23] A brutal heat wave, exacerbated by warm "loo" winds blowing south from Pakistan, exposed hundreds of millions of people to extreme heat stress. In the area around Hyderabad, five million chickens perished, and farmers faced desiccated conditions.

---

[20] www.ametsoc.org/ams/index.cfm/publications/bulletin-of-the-american-meteorological-society-bams/explaining-extreme-events-from-a-climate-perspective/.

[21] Wehner, Michael, et al. "The deadly combination of heat and humidity in India and Pakistan in summer 2015." *Bulletin of the American Meteorological Society* 97.12 (2016): S81–S86. journals.ametsoc.org/doi/10.1175/BAMS-D-16-0145.1

[22] en.wikipedia.org/wiki/2015_Indian_heat_wave.

[23] www.telegraph.co.uk/news/worldnews/asia/india/11636124/Indias-extreme-heat-wave-in-pictures.html.

Dr. Wehner et al.'s attribution study used both statistical analysis and climate models to examine the potential role played by climate change. They built a statistical model linking the amount of atmospheric CO_2 to the frequency of very warm days, showing that increasing levels of CO_2 were statistically related to increases in the frequency of heat extremes. The study then used an atmospheric model to study the frequency and magnitudes of extreme events. This atmospheric model was similar to those used to predict global weather patterns. The model was forced with observed May 2015 sea surface temperatures, sea ice extent, and atmospheric greenhouse gasses and aerosols. This process was repeated ninety-eight times, to give ninety-eight independent weather predictions. The chaotic nature of the weather ensured their independence. The India analysis team then simulated "a world that might have been" by removing estimates of human-induced changes to sea surface temperatures, sea ice distributions, greenhouse gasses, and aerosols. Such a simulated world is often called a "counterfactual" experiment representing the climate system had humans not altered the composition of the atmosphere.

As with the actual experiments, the weather model was run ninety-eight times, but forced with cooler "counterfactual" boundary conditions. So now the research team had ninety-eight sets of May 31, 2015 daily heat index values from the actual world, and ninety-eight sets of daily heat index values from a counterfactual world without climate change. Because human health impacts are much greater when we experience persistent heat stress, they next converted these daily values into running five-day averages. The figure below shows the results of these experiments. The dashed line shows the "Counterfactual" distribution of five-day heat index values in a world with climate change. The solid black line shows the "Actual" distribution from a world with climate change. Climate change made the 2015 heat waves much more probable.

The information contained in these distributions can be used to examine changes in extreme event intensity or frequency. To explore changes in the magnitude of extreme heat events similar to those experienced in 2015, the India study identified the "return period" of the event. Such events occurred about every three years

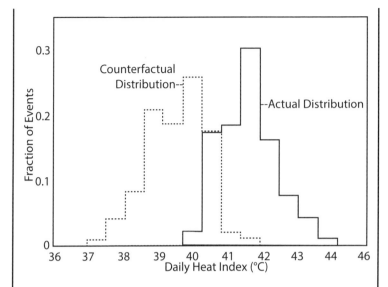

Figure 5.9 Figure showing histograms of five-day averages of simulated May 2015 daily maximum heat index values in Hyderabad for the counterfactual (dashed line) and actual (solid black line) simulations. Recreated based on results shown in Wehner et al.'s 2016 article. © American Meteorological Society. Used with permission

in the observational record. By contrasting the one-in-three year warmest heat index values in the Actual and Counterfactual distributions, Wehner et al. identified a large change in magnitude (+1.7°C). Climate change made Hyderabad heat extremes much larger. To explore changes in the frequency of the extreme heat events, the team calculated the probability of the observed peak five-day heat index value occurring in a world with and without climate change. In a world without climate change, the observed extreme event was *very* unlikely, only happening once every ninety-two years. The increase in the probability of such an event happening can be expressed as the "risk ratio." The risk ratio is calculated by dividing the probability of an event happening in a world with climate change by the probability of the event occurring in a counterfactual world without. The data presented in Figure 5.9 indicates that climate change made it ***much more likely***

(32.8 times) for Hyderabad to experience deadly heat waves similar to those that occurred in 2015. Implication? Human-induced climate change almost certainly contributed to thousands of deaths.

Sidebar: Catastrophic Fire Danger for the Greater Sydney Area

As I updated this chapter in early November 2019, catastrophic fires raged, for the first time ever, through

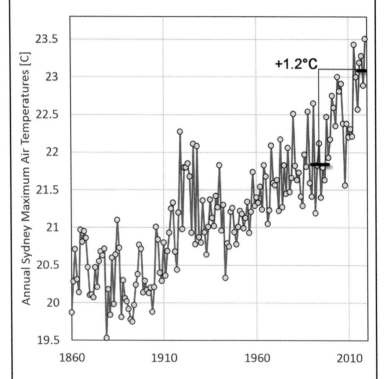

Figure 5.10 Time series of average annual Sydney daily maximum air temperatures from the Observatory Hill weather station. Data from www.bom.gov.au/climate/data/

Sydney, Australia.[24] Fueled by low rainfall levels, dry vegetation, and extremely warm temperatures, more than 120 fires burned across two states, stretching 1,000 km from Sydney to the Gold Coast near Brisbane.[25] Sydney has extensive weather data stretching back to the mid-nineteenth century, and this gives us an excellent opportunity to look at just how warm conditions have been in Sydney. Figure 5.10 shows a continuous time series of January–December annual average daily maximum air temperatures. Daily maximum air temperatures are recorded each day in the early afternoon, when the daily temperature cycle reaches its peak. Each dot in Figure 5.10 represents the average of a year's worth of these values. The year 2019 was the warmest year on record. We also note a very strong upward trend, with the 2015–2019 data more than 1.2°C warmer than the 1990s average.

---

[24] en.wikipedia.org/wiki/2019%E2%80%9320_Australian_bushfire_season.
[25] www.bbc.com/news/world-australia-50365131.

6 PRECIPITATION EXTREMES
Observations and Impacts

Introduction

On Monday, November 29, 2019, about 4:15 PM, the Cave Fire broke out two miles from my house in the woods behind Santa Barbara, within spitting distance from the Painted Cave community. Just a week prior, on Tuesday, I had discussed climate change and fire risk with my friends Ted Adams and Mike Williams on their disaster preparedness radio show. Now Ted's house was near the front lines of the flames. Thirty mile-an-hour downhill winds conspired with bone-dry brush to fuel the conflagration, which raced downhill, away from Painted Cave but toward the densely populated city of Santa Barbara. From below, Santa Barbarites watched the fire descend at insane speeds. It being winter, the sun rapidly set. With the darkness came the grounding of the fire planes, conditions being too dangerous to fly. Wind-blown embers raced downhill ahead of the walls of flames, producing a series of spot firs that helped feed the fire's rapid advance. By 8 PM the Cave Fire stretched hungry fingers toward Santa Barbara's house-filled foothills. Mandatory evacuations were called for more than 5,000 people. Throughout the night a few hundred hardy first responders stood between the city and disaster, ultimately stopping the fire and saving the city.

It just so happened that I had a prescheduled telephone call with Aurora Almendral, a *National Geographic* reporter on her way to Kenya to work on a story on climate change and migration. I stood on a hill above and behind the Santa Barbara with my wife, kids, and neighbors, watching the flames spread toward our homes, and tried to explain to Aurora how a warming atmosphere intensifies droughts, floods, and fires: *basic physics plus warming temperatures equals more extreme rainfall when the atmosphere is moist and enhanced moisture extraction from land surfaces when the atmosphere is dry.*

On December 4, 2019, the Global Carbon Budget project released their annual report.[1] In 2019, once again, about half our carbon emissions remained in the atmosphere, while about a quarter were absorbed by the ocean and a quarter by the land; carbon dioxide concentrations in the atmosphere now exceed 410 parts per million. Once again, we emitted unprecedented amounts of CO_2.

On December 18, the East African Food Security and Nutrition Working Group released a report detailing the dangerous impacts of catastrophic flooding on eastern Africa. According to rainfall observations from the Climate Hazards Center (where I work), much of the region experienced the wettest or second-wettest October–December rainy season on record (since 1981). Floods impacted more than 2.8 million individuals, directly killing 280 people and leading to widespread outbreaks of cholera and acute watery diarrhea.

In December 2019, the impact of drought and consecutive extreme heat waves also contributed to catastrophic conditions supporting ninety-eight fires in New South Wales, Australia.[2] Fires burned outside every major Australian city, not to mention additional fires in Singapore, Indonesia, and Papau New Guinea (Figure 6.1). Two volunteer Australian

---

[1] www.globalcarbonproject.org/index.htm.
[2] twitter.com/NSWRFS/status/1208717082012946432/photo/1.

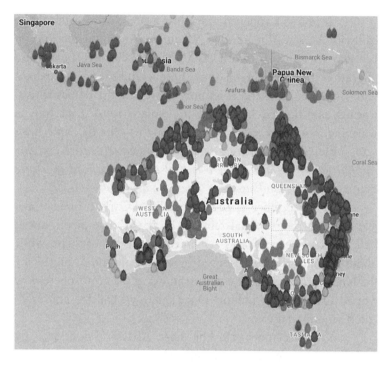

Figure 6.1 Fire locations on December 22, 2019.
Map obtained from myfirewatch.landgate.wa.gov.au/

firefighters, both young fathers (Figure 6.2), died when their truck was hit by a falling tree.[3]

A Conceptual Model of Temperature: Water Vapor Relationships

Linking all these events, and of course all life on Earth, is our atmosphere.

Our thin, thin atmosphere.

If the Earth were a basketball, the atmosphere would be 0.03 inch or 0.8 mm thick, literally whisker deep.

---

[3] www.rfs.nsw.gov.au/news-and-media/general-news/featured/support-for-firefighter-families.

125 / Precipitation Extremes: Observations and Impacts

Figure 6.2 Australian volunteer firefighters Andrew O'Dwyer (age 36) and Geoffrey Keaton (age 32) lost their lives fighting fires southwest of Sydney on December 20, 2019.

The vast majority of this atmosphere is made up of nitrogen and oxygen atoms. Nitrogen accounts for approximately 78 percent of the atmosphere, and is boring (chemically inert). Kind of like a climate scientist at a party, nitrogen just sits there without much to say. Oxygen, which makes up 21 percent or so of the atmosphere, is much more active, supporting key processes like photosynthesis and combustion. While oxygen levels have varied dramatically on geologic time scales, on climate and weather time scales they remain essentially constant.

Unlike nitrogen or oxygen, the number of atmospheric water molecules (aka water vapor) varies substantially from place to place and time to time. U these variations is critical to understanding climate change. These insights are accessible to anyone with a good comprehension of basic scientific principles. Air molecules are disconnected from each other, and free to bounce around in very random ways. A molecule, most frequently nitrogen or oxygen, will careen in some random direction until its electron cloud encounters another molecules' electron cloud, at which point they exchange momentum, promise to call each other in the morning, and head in disparate directions. The velocity of these molecules is directly related to the air's temperature and density. Cooler air with slower molecules will tend to be more dense. The molecules are packed together more closely. Warmer air with faster molecules will tend to be less dense. The molecules will be, on average, farther apart. Figure 6.3 shows this relationship schematically. Typical atmosphere molecules will usually be paired nitrogen (N_2) or oxygen (O_2) atoms. Nitrogen and oxygen make up 99 percent of the atmosphere. Slower-moving cold air parcels will tend to have smaller distances between them, making it harder for the

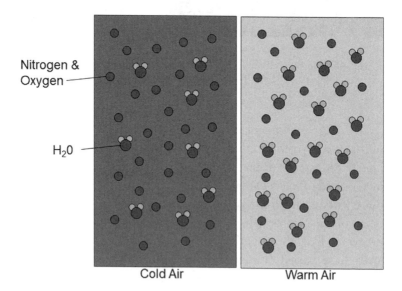

Figure 6.3 Schematic diagram of cold air (left) and warm air (right). On average, the molecules in cold air are closer together, making less room for water molecules. In warm air, the molecules are farther apart, creating more space for water molecules.

air to hold water vapor molecules (shown as a big oxygen atom with two small oydrogen atoms attached). Warming the air causes the nitrogen and oxygen to move more quickly and spread out, creating more space for water vapor.

Ironically, the fact that a warmer atmosphere can hold more water vapor can lead to increased drought intensities and drier fire fuels. At small scales near the surface of a plant or a clod of earth, water vapor molecules may move up from the surface into the air, but this process can slow or even stop completely depending on the air's temperature and total amount of preexisting water vapor in the lowest level of the atmosphere. If we keep adding water vapor to any parcel of air at any temperature, it will eventually become "saturated, with a relative humidity of 100 percent." The parcel is no longer able to absorb more water vapor. But warmer air can hold a lot more water vapor; increasing the temperature by one degree Celsius increases the potential water-holding capacity of air by about 7 percent. This relationship is commonly referred to as the Clausius-Clapeyron relationship. Under dry conditions, like during the gap between one rainy season and the next, increases in air temperatures will often lead to air that is less saturated, making it easier for vegetation to lose moisture. A modest level of warming, supporting modest enhancements of plant water loss month after month, can produce substantial decreases in plant and fuel moisture.

Decreases in plant and fuel moisture can in turn lead to increased fire risks in, for example, places like California and Australia. Greater vegetative moisture loss can also dry out grasslands, providing less food for cattle, sheep, goats and camels. These moisture losses have large impacts on both lucrative ranches in the US[4] and on poor, food-insecure pastoralists in Africa.[5]

---

[4] Williams, Emily, et al. "Quantifying human-induced temperature impacts on the 2018 United States Four Corners Hydrologic and Agro-Pastoral Drought." *Bulletin of the American Meteorological Society* 101.1 (2020): S11–S16.
   https://journals.ametsoc.org/doi/pdf/10.1175/BAMS-D-19-0187.1.
[5] Pricope, Narcisa G., et al. "The climate-population nexus in the East African Horn: Emerging degradation trends in rangeland and pastoral livelihood zones." *Global Environmental Change* 23.6 (2013): 1525–1541.
   https://www.sciencedirect.com/science/article/pii/S0959378013001738

But the same simple physical principle illustrated in Figure 6.3 (the Clausius-Clapeyron relationship) can also lead to increased precipitation extremes. Nitrogen and oxygen molecules are very well mixed in our atmosphere. Everywhere you go, there they are. Not so with water vapor. Warmer air holds more water vapor, so we usually find more water vapor in the tropics. But even at a fixed latitude we will see huge variations over time and place. When air is close to saturation (100 percent relative humidity), warming increases the total amount of water available for precipitation.

A simple conceptual model of rainfall (Figure 6.4) can help us understand why climate change is making extreme precipitation even more extreme. Heavy precipitation occurs when moist air rises quickly into the upper atmosphere and cools. It gets colder as you go up in the atmosphere because the air becomes less dense. This cooling reduces the air's ability to hold water vapor. When the rising parcel of air becomes saturated, the water vapor it contains condenses, becoming liquid droplets or ice crystals: clouds and precipitation. Warmer air will contain more water vapor, supporting heavier rainfall.

A little algebra can help us express this mathematically. We can describe rainfall rates as the multiplicative interaction of two terms – water vapor (W) and the vertical velocity (V). W is simply the amount of gaseous water vapor in a parcel of air. V is the vertical (upward when raining) velocity of that parcel. Rain happens when moist air moves up and cools, causing the water vapor in the air to cool, condense, and precipitate. So we can assume that rain rates (RR) will be proportional to W times V (RR = W × V) · W × V describe the vertical mass flux of water, which is strongly related to precipitation intensities (i.e., millimeters or inches of rainfall per day).

Now, if we do a little thought experiment (Figure 6.4), we can relate this simple precipitation model to increases in air temperature. On the bottom left and bottom right we have parcels of air with the same relative humidity, but the right-hand parcel is warmer. It therefore has substantially more water vapor molecules. Now we subject both parcels to the same vertical velocity,

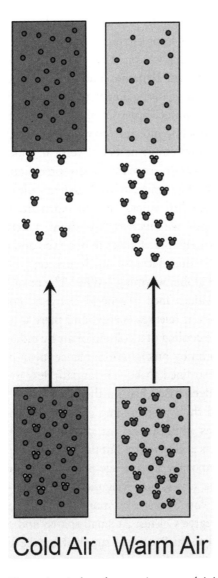

Figure 6.4 A thought experiment explaining the impact of increased water vapor on precipitation rates.

whisking them up into an ever-colder atmosphere. Before long, the water in each parcel will condense. Assume the velocity (V) is the same in both cases. Since the warmer parcel holds more water vapor, it would likely produce more precipitate. W would be bigger. This simple relationship provides a good starting point for understanding changes in precipitation extremes.

Climate change impacts on these two terms, W and V, are generally called thermodynamic and dynamic[6] controls. Thermodynamic control is just a fancy term to describe how warmer air holds more water vapor, i.e. the Clausius-Clapeyron relationship. As atmospheric moisture content increases, precipitation increases.[7] For our simple model, if vertical velocity stays the same but water vapor increases, then precipitation will increase in proportion to increases in W. A general rule of thumb is that if relative humidity values remain constant, then the amount of water vapor will increase by about 7 percent per degree Celsius of warming. But changes in observed and modeled *average* precipitation values increase much less, only about 3.4 percent per degree of global warming.[8] Why? The answer is that mean precipitation values face a general energetic constraint; when water condenses, it releases energy, and there is a balance of energy between the cooling of a column of air by radiation and the heating of a column by precipitation. Hence mean precipitation changes are constrained to be substantially less than those expected due to changes based just on the relationship between air temperatures and the water holding capacity of a warming atmosphere. This has implications for droughts, because our warming atmosphere's ability to dry out the land may be increasing faster than the atmosphere's average precipitation rates.

At short time scales, however, thermodynamic controls on rainfall extremes can increase by more than 3.4 percent or even 7 percent per degree Celsius. At small spatial and temporal scales, the "thermodynamic" control supplied by the Clausius-Clapeyron relationship predicts a ~7 percent increase in heavy precipitation rates per degree of warming. As we see in Figure 6.4,

---

[6] Emori, Seita, and S. J. Brown. "Dynamic and thermodynamic changes in mean and extreme precipitation under changed climate." *Geophysical Research Letters* 32.17 (2005).

[7] Trenberth, K. E., A. Dai, R. M. Rasmussen, and D. B. Parsons, "The changing character of precipitation." *Bulletin of the American Meteorological Society* 84 (2003): 1205–1217.

[8] Allen, Myles R., and William J. Ingram. "Constraints on future changes in climate and the hydrologic cycle." *Nature* 419.6903 (2002): 224.

a warmer atmosphere is like a bigger sponge. When this sponge rises and cools, more water can condense and fall as rain.

But as we will see when discussing cyclones and hurricanes, warming impacts on the most extreme rainfall events could become even more severe. In extreme rainfall events moisture doesn't just move up; it also moves side to side. Superstorms like Hurricane Harvey draw in winds from hundreds of miles away, collecting additional moisture, to potentially explosive effect. This extra moisture increases rainfall rates. But condensation also releases a tremendous amount of energy, potentially creating a weather feedback that leads to more rapid moisture convergence and accelerated vertical motions.

Examining the Data

While a detailed analysis is beyond the scope of this book, I always want to present you with real data to underscore the topics we are discussing. Here we briefly examine five-day precipitation totals from the archive of the Climate Hazards Center for Infrared Precipitation with Stations (CHIRPS). This is the precipitation product produced by our group (the Climate Hazards Center, CHC) at the University of California, Santa Barbara.[9] My friend Pete Peterson and other members of our team have spent years making this one of the best global rainfall products. CHIRPS combines rainfall estimates from geostationary infrared weather satellites with observed rain gauge observations. This rapidly updated product was designed to support humanitarian relief efforts. Here we analyze global CHIRPS data following an approach commonly used to examine changes in extremes.[10] For each year and location, an annual maximum rainfall accumulation total is identified, and then divided by that

---

[9] chc.ucsb.edu.
[10] Donat, Markus G., et al. "More extreme precipitation in the world's dry and wet regions." *Nature Climate Change* 6.5 (2016): 508; Donat, Markus G., Oliver Angélil, and Anna M. Ukkola. "Intensification of precipitation extremes in the world's humid and water-limited regions." *Environmental Research Letters* 14.6 (2019): 065003.iopscience.iop.org/article/10.1088/1748-9326/ab1c8e#erlab1c8ef1.

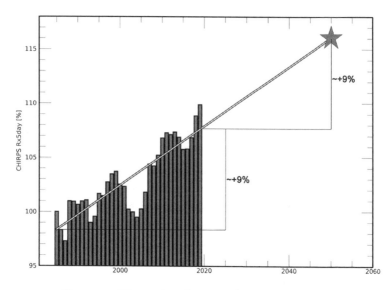

Figure 6.5 Time series of running five-year averages of global five-day terrestrial extreme precipitation rainfall totals expressed as percentages of the 1981–1985 value. Also shown is a line describing the data's strong linear trend along with a 2050 projected value.

location's long-term average total annual precipitation. This is a measure of the intensity of each location's most extreme precipitation for that year. We can then calculate five-year global averages of these values to look for trends in extreme precipitation. To facilitate analyzing changes over time, Figure 6.5 displays these values expressed as the percentages of the first 1981–1985 five-year value. While there are natural year-to-year fluctuations, we also a see a strong upward trend in the data. Between the first five-year period (1981–1985) and the last (2015–2019), we see about a 9 percent increase in the observed extreme rainfall events.

A 9 percent difference may not sound like much, but it can really be substantial, increasing the intensity of very extreme rainfall events. Flooding is all about the *extra* water. Runoff is what we call the water that is left over after rainfall is absorbed by the soil and evaporated up into the atmosphere. In simple terms, Runoff = Rainfall − Evaporation. At low precipitation intensities, the Sink can be as big as the Source. This relationship tends to change in a nonlinear way when a watershed receives a lot of water, so a 9 percent increase in a rainfall extreme may increase runoff (and flooding) by much more. If the observed trend persists, we will likely see another large increase in the wettest five-day precipitation events (star in Figure 6.5). This is a recurring theme in this book. We have already seen a substantial increase in precipitation-related hazards that are harming people today. More increases are on the way. While the atmosphere is a complicated place, the relationship between atmospheric warming, increases in water vapor, and associated enhancements to extreme precipitation are basic physics that everyone can understand (Figures 6.3 and 6.4).

2015–2019 Impacts of Extreme Precipitation

While we will deal with hurricanes and cyclones in the next chapter, here we will briefly discuss some of the most severe floods and storms that happened between 2015 and 2019. Between 1998 and 2017, floods and storms (including the impact of hurricanes) affected more people than any other type of disaster, impacting 2.7 *billion* people overall and resulting in 1.99 *trillion* dollars of recorded economic losses.[11] Most of the wettest extreme rainfall events tend to occur in Asia and the Maritime Continent (the region of Southeast Asia that comprises, among other countries, Indonesia, Philippines, and Papua New Guinea), with heavy precipitation also occurring in northern Australia, southeast Africa and Madagascar, South America, Central America, and the southeastern United States. We can explore the potential impact of the extremes by digging into the international emergencies database (EM-DAT), which is maintained by the Centre for Research on the Epidemiology of disasters (www.cred.be/). This is the database used by the United Nations Office for Disaster Risk Reduction. Annual

---

[11] In 2017 US dollars. United Nations report, Economic Losses, Poverty & Disasters: 1998–2017, www.unisdr.org/files/61119_credeconomiclosses.pdf.

disaster statistical reviews and other interesting analyses are available on the CRED site (www.cred.be/publications). According to EM-DAT, in 2015–2019, severe storms (including cyclones, typhoons, and hurricanes), floods, and landslides have affected some 405 million people, resulting in more than 31,000 deaths and $569,437,073,000 ($569 *billion*!) in damages. While quantifying such impacts with great precision is difficult, it is clear that flooding and rainfall extremes are expensive and dangerous.

Focusing just on large flood events, we can explore the data in more detail. Table 6.1 summarizes recent (2015–2019) EM-DAT[12] disaster statistics for big flood events. Big flood events are defined as those impacting at least 1.5 million people. As you can see, most of these big-impact events occur in Asia. This spatial selection arises from an interaction between human development and natural climate variability. This is the most densely populated region on Earth. But it also contains many regions that customarily receive abundant and sometimes extreme rains. Just over five years, the total number of people being affected is pretty staggering: 223 million. Associated with these floods were a total of 9,036 deaths and economic losses of approximately $78 billion.

To put a human face on these statistics, we can consider some of the worst flood disasters from 2018. A deadly flood occurred in the Indian state of Kerala, which is on the western side of the Ghats Mountains in southwestern India. The western side of the Ghats is one of the rainiest places on Earth, because warm moist monsoon winds blow right against the mountain range. This produces strong vertical motions. Strong vertical motions combined with high levels of water vapor produces intense rainfall events ($RR = W \times V$). In Kerala in 2018 torrential rains quickly accumulated, dams needed to be opened to release the water, and more than 500 people are believed to have directly died. Extensive inundation displaced more than a million people and impacted more than 23 million people, causing

[12] www.emdat.be/

Precipitation Extremes: Observations and Impacts

Table 6.1. EM-DAT disaster statistics for non-cyclone 2015–2019 floods affecting at least 1.5 million people.

Year	Country Name	Total Deaths	Total Affected	Total Damage (Millions US$)
2016	China	816	61,297,200	31,793
2019	India	1752	23,830,060	unknown
2018	India	710	23,307,698	2,865
2017	India	1046	22,271,843	2,117
2015	India	839	16,413,459	2,880
2017	China	231	12,333,608	8,446
2019	Iran	75	10,001,096	2,580
2017	Bangladesh	144	8,086,025	628
2019	China	152	6,375,861	unknown
2019	Bangladesh	119	4,000,000	unknown
2017	Thailand	150	3,790,498	1,308
2018	China	244	3,698,400	4,570
2016	Dominican Republic	15	2,792,000	unknown
2016	Philippines	45	2,563,098	9
2017	Peru	200	2,188,505	3,200
2018	Nigeria	300	1,938,204	275
2016	Bangladesh	106	1,900,000	150
2017	Philippines	22	1,842,000	8
2016	India	666	3,806,000	1,499
2017	Nepal	187	1,706,134	595
2015	Myanmar	172	1,635,703	119
2015	Pakistan	367	1,577,490	1
2018	Japan	246	1,500,102	9,500
2016	Viet Nam	91	1,428,585	161
2015	Bangladesh	31	1,411,901	40

Table 6.1. cont'd

Year	Country Name	Total Deaths	Total Affected	Total Damage (Millions US$)
2015	China	310	1,374,350	6,992
2017	Ghana		1,000,000	unknown
Total		9,036	223,069,820	79,737

Source: EM-DAT: The Emergency Events Database – Universite catholique de Louvain (UCL) – CRED, D. Guha-Sapir – www.emdat.be, Brussels, Belgium. Empty cells indicate missing values. Damage amounts present reported values in US dollar equivalents from the year of the disaster. Created December 23, 2019.

an estimated $4 billion in damages and leading to at least 319 fatalities.

Let's look at another example. The summer storms that struck Japan between June 28 and July 9 in 2018 were even more financially devastating. On June 28, heavy rains began falling across southwestern Japan. On July 3, during Typhoon Prapiroon, moist air and an extra-tropical low-pressure system combined to produce very intense rains. Daily rainfall totals exceeded 250 millimeters. In some places, eleven-day rainfall totals quickly exceeded 1,850 millimeters (Figure 6.6). The intensity of these totals is amazing. In the village Umaji of Umaji, in Kōchi prefecture, four consecutive days received rainfall totals exceeding 250 millimeters. A 1 meter × 1 meter × 1,850 millimeter volume of water would weigh 1,853 kilograms. Japan reeled from the impact of this flooding, which contributed to a ~1.2 percent contraction of the Japanese economy.[13] According to EM-DAT, this deadly natural disaster resulted in 246 mortalities and a *billion* dollars in economic losses.

Climate Change Projections

Are we heading toward more such extreme precipitation events? To explore this question we can use a set of

[13] www.bbc.com/news/business-46203864.

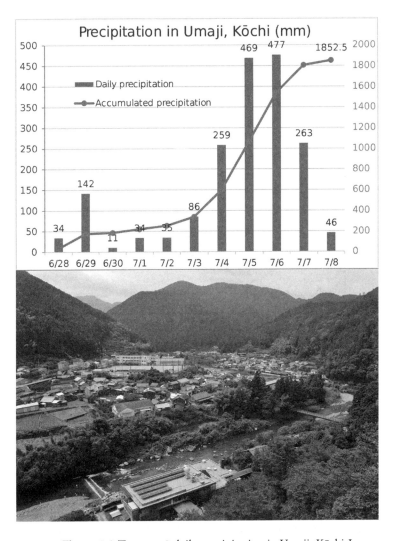

Figure 6.6 Top: 2018 daily precipitation in Umaji, Kōchi Japan. Bottom: A picture of Umaji taken by Saigen Jiro.

annual simulated climate extreme statistics hosted at the Koninklijk Nederlands Meteorologisch Instituut (the Dutch national weather service) on the great Climate Explorer website (climexp.knmi.nl). These climate extremes were obtained from the Canadian Centre for Climate Modeling

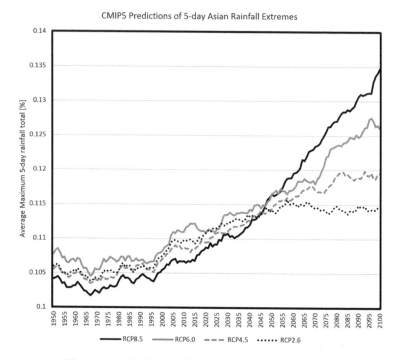

Figure 6.7 CMIP5 five-day precipitation extremes, for Asia and Oceania, expressed as a fraction of the average annual precipitation.

and Analysis[14] using a large multi-model ensemble of climate change simulations.[15] Because so many extreme precipitation events happen over Asia and Oceania, I have calculated averages over land between 65°E and 152°E and 20°S and 40°N. As in Figure 6.5, the annual maximum five-day precipitation total is expressed as a fraction of the 1950–1979 mean annual total precipitation. Four time series are presented in Figure 6.7. If we continue along our current rapid emissions scenario (RCP8.5), we are likely to see continued rapid increases in precipitation, with the models anticipating that typical five-day maximum rainfall totals will be about

---

[14] climate-modelling.canada.ca/climatemodeldata/climdex/climdex.shtml.
[15] These results are based on the Phase 5 Coupled Model Intercomparison Project (CMIP5) ensemble.

35 percent more intense than they were in the 1950s. Given that we are already experiencing large increases in rainfall extremes with impacts costing hundreds of billions of dollars, resulting in serious disruptions in development,[16] continued increases in extreme precipitation will be costly and deadly.

Note, however, that even modest reductions in greenhouse gas emissions, such as those associated with the RCP4.5 and 6.0 pathways, really make a big difference by mid-century. According to these simulation results, Asian precipitation extremes increased by about 5 percent between 1950 and 2018. If we act now to curb emissions, we may see a similar increase by mid-century. Following a rapid growth (RCP8.5) trajectory means that we may see rainfall extremes increasing twice as fast. The decision on which path we take is being made now; whether we *choose* to decide or not, we still have made a choice.[17] There is solid observational (Figure 6.5) and model-based (Figure 6.7) evidence supporting the link between a warming atmosphere and more intense precipitation extremes, and clear evidence that these extremes are having deadly and costly impacts *today* (Table 6.1). Relatively modest (RCP4.5) or aggressive (RCP2.6) reductions in emissions will limit increases in these human catastrophes and attendant human suffering.

---

[16] gar.unisdr.org/sites/default/files/gar19distilled.pdf.
[17] Based on the band Rush's song, *Freewill*.

7 HURRICANES, CYCLONES, AND TYPHOONS

Solon, the ancient Greek philosopher famed for giving the city of Athens its laws, famously told the rich, successful, and powerful King of Lydia, "Count no man lucky until his end is known"[1] (Figure 7.1). How true. Each of us, no matter how surrounded by kindly kin, treasure, glory, and the good opinion of others, could be wracked with ruin in an instant. Failing brakes, the cough of a sick child, cancer. So fragile we are. So intertwined with other. So ultimately enmeshed in the universe around us. This is a good thing, our existential thingness, which gives us an opportunity to be *good*. And that goodness knows that the potential for catastrophe – drought, flood, fire – links us in a common web of humanity.

Life tends to be conservative because there are so many ways to not be it (alive). Between the ultimately cool and absolutely hot, life hangs in a delicate balance. We live in a narrow temperature band, and as climate change shifts our planet's narrow temperature regime, we may experience more extreme hurricanes and cyclones, which often bring catastrophic damage.

At absolute zero – zero degrees Kelvin or −273.15 degrees Celsius – the motion of molecules ceases. This is the

[1] classicalwisdom.com/philosophy/count-no-man-happy-end-known/.

Figure 7.1 Solon before Croesus, by the Dutch painter Gerard van Honthorst, painted in 1624. https://en.wikipedia.org/wiki/Gerard_van_Honthorst#/media/File:Honthorst_solon_and_croesus.jpg

absolute minimum temperature, absolute cold. Then there is absolute hot, about which scientists are somewhat unsure. According to quantum theory derived in the late 1980s, the maximum obtainable temperature would be the Planck temperature, or $\sim 1.416808c \times c10^{32}$ Kelvin. But let's be conservative, and dial back to the temperature of a supernova, which is about 100 billion Kelvin. So temperatures range from ~ -273.15C to $\sim 10^{11}$°C, or ~ 0 to $100{,}000{,}000{,}000$ Kelvin.

Between these extremes, life hovers. Let's sketch out the potential range of temperatures on Earth, beginning at the hypothetical coolest: Miles Davis's Sextet playing "Kind of Blue" in Antarctica (-98°C or 175K). And then for hot, you pretty much can't beat the emission temperatures (~ 50°C or ~ 323K) coming from the parking lot of the Devil's Tower National Monument in northeastern Wyoming in August, during the annual Sturgis Motorcycle Rally.[2] While to me the Antarctica-to-Sturgis spread seems exciting, exotic, and better than Monday, from a galactic scale we are looking at mere digits, about 200K in a universe that ranges from 0 to ~10^{11}K. Pessimists should observe that we have a lot more room on the upside, where global warming abides.

Of course, those perceptive souls who might wish to believe in a benign universe or caring higher power *could* mention that we seem to be luckily ensconced within a few magic decimal places conducive to watery phases and associated life-sustaining metabolic processes. Out of this many decimal places, we happened happily in the two bold ones a few places from the rightmost digit: 10000000 **00** 00.

Maybe it is totally random that our planet has gyrated for billions of years, almost (in cosmic age) since the birth of the universe, between temperatures supporting freezing liquid and gaseous water. On the other hand, if I were playing God, and looking forward to the imminent and immanent evolution of human life, I would set the temperature dial to that small fraction of the spectrum between ~175 and 323 Kelvin (~−98 to +50°C). And there we are. Water, in its myriad of forms, makes complex life possible on our Blue Planet (Figure 7.2).

Our Blue Marble's temperature range uniquely (at least as we so far know) supports water in all its three glorious phases (solid, liquid, gas). Near your house there might be a nice cool pond or lake or river. Above this liquid water body you will find moist air. In this moist air, water vapor (water in its gaseous state) freely mixes with our atmosphere's more boring constituents, like nitrogen and oxygen. Imagine, unseen to you, a raging house party, with all those wild and crazy

---

[2] Please note the following from www.sturgismotorcyclerally.com/faq-and-stats/ on December 26, 2018: "To be married in Sturgis® (Meade County) you must obtain a marriage license at the Register of Deeds office located in the Erskine Building at 1300 Sherman St., Suite 138. (1 block south of Main St.) Both applicants must be present with identification and $40 in cash or travelers' checks. There is no waiting period, the license is good for (20) days and same day marriages are legal. You are responsible for locating your own wedding official."

Figure 7.2 Our Blue Marble. A view of most of North America taken from a low orbit of about 826 km altitude. This composite image uses a number of swaths of the Earth's surface taken on January 4, 2012. By NASA/NOAA/GSFC/Suomi NPP/VIIRS/Norman Kuring, www.nasa.gov/multimedia/imagegallery/image_feature_2159.html.

H_2Os bouncing around. Folks in the pond would like to come to the party, but there is only so much room. The room is packed. If someone comes in the door, someone falls out the window. Unless air temperatures increase. Then the spaces between the molecules of nitrogen and oxygen increase, making more room to party.

Now assume our pounding party pad is pendulous, suspended beneath a giant hot air balloon. As we drift higher, air temperatures quickly drop. The warmth of the air around us is strongly influenced by the weight of the air above, packing the atmospheric molecules closer together. The lowest 85 percent of the atmosphere (the troposphere) stretches up about 12 kilometers where it reaches the "tropopause." The tropopause region is where incoming radiation causes the very thin air of the upper atmosphere to warm. As temperatures cool, there is less room in the "inn." As our pendulous party pad ascends, declining air temperatures quickly cool our mini-mighty party partners. They stop to check their disco-watches (digital of course) and realize it's time to chill.

Now something important happens. Slowing down, they aggregate, undergoing a "phase transition" – condensing into water droplets. These droplets, denser than the surrounding air, drop – i.e., precipitate. As our pendulous party pad rises higher, the disco palace reaches freezing temperatures. Panicked partiers leap from windows, freeze together in clumps, and drift slowly down to Earth. None of us is truly alone. Even snowflakes are compound creatures.

Precipitation processes are super-critical to supporting life on our planet. Sometimes, though, we can get way too much of a good thing. Increasing air temperatures can contribute to more extreme hurricanes and floods. As air temperatures increase, the water-holding capacity of the atmosphere goes up in a predictable systematic way, as we saw in the previous chapter. This additional water vapor is expected to directly increase the intensity of extreme precipitation. An increase in the size of our pendulous party pad means more precipitating panicking partiers.

But there are also important indirect dynamic effects. Water phase transitions can release or absorb a lot of energy. It takes a lot of energy to boil or freeze liquid water. When gaseous water vapor turns liquid or freezes, a tremendous amount of energy can be released – which can have important dynamic impacts. This released energy can rapidly heat the atmosphere, causing rapidly rising columns of air. Which in turn causes precipitation, triggering precipitating partying panickers that release more energy – a positive feedback loop.

The "collective" aspect of atmospheric water vapor exacerbates both of the direct (7 percent per degree) and indirect dynamic effects of water vapor increases. Unlike atmospheric temperatures and pressures, which tend to be evenly distributed in space and time, water vapor can vary greatly from place to place in the atmosphere as winds collect converging flows. Sweeping across warm waters, winds can gather water from a vast fetch, sometimes to catastrophic effect.

Take, for example, Hurricane Harvey, which pummeled the coast of Texas and Louisiana in late August 2017. The greater Houston area received 30 to 40-plus inches of rain, with several totals well above 50 inches (127 cm) of rain from Harvey,[3] making it the largest rainfall event on record in the lower forty-eight states. As reported in the *Washington Post*,[4] John Nielsen-Gammon, Texas state climatologist, found that "Harvey's total rainfall concentrated over a 20,000-square-mile area represents nearly 19 times the daily discharge of the Mississippi River." According to the *Post*, this is equivalent to about 9 *trillion* gallons of water,[5] equivalent to a cube of water two miles long, two miles wide, and two miles high; this would be enough water to fill the Great Salt Lake in Nevada, twice. In Houston, more than five million people lived in areas receiving more than 36 inches of rain.[6] More than 32,000 people were forced to abandon their homes and sleep in shelters.[7] Something to always remember, however, is that water isn't just wet – it's also energy. When 9 trillion gallons of water are evaporated from the ocean, the ocean is cooled, and a *lot* of energy is stored up in the atmospheric water vapor. The total energy associated

---

[3] www.weather.gov/hgx/hurricaneharvey.
[4] www.washingtonpost.com/news/capital-weather-gang/wp/2017/08/29/harvey-marks-the-most-extreme-rain-event-in-u-s-history/?utm_term=.3a6689ffe5b3.
[5] www.washingtonpost.com/news/capital-weather-gang/wp/2017/08/27/texas-flood-disaster-harvey-has-unloaded-9-trillion-tons-of-water/?tid=a_inl&utm_term=.066934a16c9d.
[6] https://mashable.com/2017/08/29/harvey-houston-flood-by-the-numbers-worst-flood/.
[7] www.npr.org/sections/thetwo-way/2017/08/30/547227788/harvey-makes-landfall-again-in-louisiana.

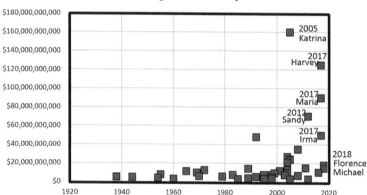

Figure 7.3 The most expensive US hurricanes, expressed in terms of adjusted 2017 US dollars. Based on data provided NOAA's National Hurricane Center.

with evaporating 9 billion gallons of water is about 76×10^{18} Joules (76 quintillion Joules; I had to look that one up). When water vapor condenses during precipitation, this same amount of energy is released. This energy rapidly warms the atmosphere, fueling a hurricane's gale-force winds. Over Texas, peak wind gust speeds reached 132 miles (211 km) per hour. In 2017, all the humans around the world consumed about 630 quintillion Joules of energy – or just 8.3 times the amount of energy released in Texas by Hurricane Harvey.

According to NOAA,[8] the economic impacts of Harvey were massive – $125 billion, making it the second most expensive hurricane after Katrina (Figure 7.3). Note also that two other very expensive hurricanes occurred in 2017: Maria, which struck Puerto Rico and the Virgin Islands; and Irma, which struck Florida. Maria devastated Puerto Rico,[9] contributing to the deaths of at least 2,975 people and leaving the island without power for months. Irma was the strongest observed hurricane in the Open Atlantic, causing catastrophic damage to Florida and the

[8] www.nhc.noaa.gov/news/UpdatedCostliest.pdf.
[9] www.nationalgeographic.com/magazine/2018/03/puerto-rico-after-hurricane-maria-dispatches/.

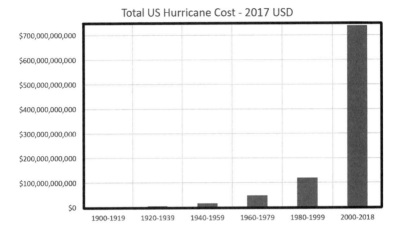

Figure 7.4 Total cost of the biggest hurricanes summed by twenty-year time periods.

Caribbean Islands along its path. In 2018, hurricanes Florence and Michael struck the southeastern United States, causing some $33 million in damages. These three cyclones (Harvey, Maria, Irma) alone inflicted some $260 billion in damages, according to the NOAA data shown below (Table 7.1). To put this in perspective, the 2003–2010 US–Iraq war cost US taxpayers about $822 billion in direct military expenditures, according to the November 2018 Brown University Cost of War report.[10]

We can rearrange the data from Figure 7.3 by summing up impacts over twenty-year time periods (Figure 7.4). This emphasizes the huge jump in economic costs incurred by the United States since the beginning of the twenty-first century. I'm in the disaster business, and used to looking at bad news, but even I was shocked by Figure 7.4. Together, the biggest hurricanes inflicted about $740 billion in damages between 2000 and 2018. So, so far, twenty-first-century hurricanes have been almost as costly as the Iraq War. There are many factors related to these cost increases, and a very important contributing factor is that people like to live next to the ocean. According to the

---

[10] watson.brown.edu/costsofwar/files/cow/imce/papers/2018/Crawford_Costs%20of%20War%20Estimates%20Through%20FY2019.pdf.

Wall Street Journal,[11] between 1980 and 2017, population density in coastal areas of the US Gulf and eastern states increased from about 350 to 550 people per square mile, while densities in the rest of the US remained very steady. And coastal real estate and repairs are expensive. While there are debates regarding the true inflation-adjusted costs of these extremes, most assessments indicate that, indeed, there has been a very large recent increase in US hurricane-related losses.[12]

Complexities Surrounding Cyclone Attribution and Detection

It is important to note, however, that increasing catastrophic losses do not *necessarily* implicate climate change. The relative complexity and rarity of hurricanes and cyclones makes their detection and attribution very difficult. According to an authoritative National Academies report, "The frequencies and intensities of tropical cyclones and severe convective storms are related to large-scale climate parameters whose relationships to climate change are understood to varying degrees but, in general, are more complex and less direct than are changes in either temperature or water vapor alone."[13] Luckily, for this chapter we can draw on two recent multiauthored synthesis reports prepared by the World Meteorological Organization Task Team on Tropical Cyclones and Climate Change. The first report focuses on observed cyclone detection and attribution results.[14] The second study[15] assesses model-projected changes in tropical cyclone activity for a 2°C anthropogenic warming.

---

[11] www.wsj.com/articles/the-rising-costs-of-hurricanes-1538222400.
[12] www.pnas.org/content/116/48/23942.short.
[13] www.nap.edu/catalog/21852/attribution-of-extreme-weather-events-in-the-context-of-climate-change.
[14] Knutson, Thomas, et al. "Tropical cyclones and climate change assessment: Part I. Detection and Attribution." *Bulletin of the American Meteorological Society* 2019 (2019). journals.ametsoc.org/doi/pdf/10.1175/BAMS-D-18-0189.1.
[15] Knutson, Thomas, et al. "Tropical cyclones and climate change assessment: Part II. Projected response to anthropogenic warming." *Bulletin of the American Meteorological Society* 2019 (2019). journals.ametsoc.org/doi/pdf/10.1175/BAMS-D-18-0194.1.

To understand these studies we need to expand our arsenal of understanding to encompass two sets of terms: detection and attribution; and type I and type II errors. In climate change assessments, according to the IPCC, "**Detection** of change is defined as the process of demonstrating that climate or a system affected by climate has changed in some defined statistical sense without providing a reason for that change," while **Attribution** is "the process of evaluating the relative contributions of multiple causal factors to a change or an event with an assignment of statistical confidence."[16] Some aspects of climate change are much easier to detect than others. At both global and local scales, for example, it is generally quite possible to detect increases in temperature extremes. In Chapter 5 we assessed the global variations of heat extremes, where we found a strong detectable signal, but even when we zoomed in to a local weather station on Observatory Hill in Australia, we found a clearly visible increase in air temperatures. Such increases are much harder to see in extreme precipitation records and in time series of data associated with cyclones. These very rare and complicated events have very limited data, which results in small sample sizes, which can make detection and trend analysis difficult.

Which brings us to the idea of type I and type II errors. These kinds of errors are often discussed in the context of medicine or legal matters. Type I errors involve accepting something as true when it is not. Type II errors involve incorrectly rejecting a hypothesis when it is true. Different analytical approaches tend to favor one type of error over the other. As Elisabeth A. Lloyd and Naomi Oreskes point out,[17] the most common approaches used by climate scientists tend to minimize type I errors. While rigorous, these approaches risk underestimating the role of global climate change in extreme events and missing connections that are really there. A series of recent

---

[16] https://www.ipcc.ch/sr15/chapter/glossary/.
[17] Lloyd, Elisabeth A., and Naomi Oreskes. "Climate change attribution: When is it appropriate to accept new methods?" *Earth's Future* 6.3 (2018): 311–325. doi.org/10.1002/2017EF000665

"storyline"-based studies focus on minimizing type II risks. According to LLoyd and Oreskes, "The storyline method is like an autopsy: it gives an account of the causes of the extreme event – the flood or storm – and can indicate whether climate change was one of these causes." The approach, however, may overstate the role of climate change.

Lloyd and Oreskes argue that climate scientists should be willing to look at climate extremes through both lenses, and this is what the recent Task Team assessment of observed cyclone changes does. In general, cyclone data tends to show relatively few clear, well-understood trends. The Team assessment finds little clear evidence, in the observations, for changes in cyclone frequency, intensity, or number of landfalls. Such evidence would involve being quite sure that such a result was unusual when compared to natural variability.

However, when viewed from a type II (risk avoidance) perspective, a large proportion of the team members agreed on the following:

- There has been a detectable observed northern migration of the latitude of maximum intensity in northwest Pacific basin, and anthropogenic forcing has contributed to this movement.
- There has been a detectable increase in the global proportion of tropical cyclones reaching category 4 or 5 intensity in recent decades, and anthropogenic forcing has contributed to this increase.
- There has been a detectable increase in the global average intensity of the strongest tropical cyclones since the early 1980s, and anthropogenic forcing has contributed to this increase.
- There has been a detectable long-term increase in the occurrence of Hurricane Harvey–like extreme precipitation events in the Texas region, and anthropogenic forcing has contributed to this increase.

Interestingly, the type II framework also yielded some negative findings. All the authors, for example, agreed that there has not been a detectable increase in the frequency of land-falling hurricane frequency since the late 1800s. So what emerges from this study is that while it is still hard to detect changes in hurricane behavior with a high degree of certainty, the balance of evidence does suggest that the strongest storms may be becoming more intense and associated with heavier precipitation. These results are broadly consistent with the Team's +2°C model-based analysis, which found that with "medium-to-high confidence projections include increased tropical cyclone rainfall rates, intensity, and proportion of storms that reach Category 4–5 intensity globally."[15]

Putting this all in layperson's terms, we might summarize these results as: to date, things look concerning, and while it is hard to be certain, model projections and the balance of observational evidence converge on an increased frequency of very strong and very rainy hurricanes and cyclones. We have clear evidence of the increasing cost of extreme hurricanes and cyclones (Figure 7.4). Using the Munich Re Catastrophe Database, we can also see a large increase in the number of hydrologic disasters (Figure 7.5). According to the database, the number of expensive ($25 million or more in 2014 US dollars) hydrologic disasters has increased from around 100 per year in the early 1990s to almost 400 per year in the 2015–2019 period. This increase is only partly due to climate change; the expansion of human settlements, especially along the coasts, puts more people in harm's way. But the rapid increase in risk is irrefutable.

It doesn't take a cyclone to wreak havoc. Many floods and disasters, in fact, are associated with non-cyclonic extreme precipitation events. In 2017, numerous catastrophic flood events occurred, most frequently in Asia.[18] Approximately 27 million people were affected by severe flooding in India, Nepal, and Bangladesh. Extensive August rains in upstream areas of Nepal and India helped fuel severe flooding in Bangladesh, with the International Federation of Red Cross and Red Crescent Societies estimating that 41 million people were affected, leading to the loss of some 950,000 homes and 1,200 lives.[19] Then in late

---

[18] 2017 Annual Disaster Statistical Review, cred.be/sites/default/files/adsr_2017.pdf
[19] www.cnn.com/2017/09/01/asia/bangladesh-south-asia-floods/index.html.

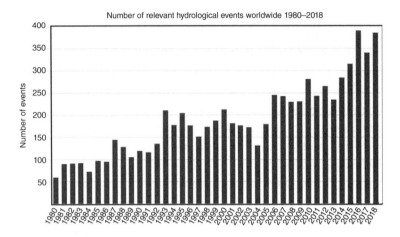

Figure 7.5 The number of large hydrologic disasters from the Munich Re catastrophe database.
Source: natcatservice.munichre.com/

June and early July 2018, successive extreme rainfall events struck Japan, triggering landslides and massive flooding. Some 158 Japanese people lost their lives, and costs to repair infrastructure are estimated at around 270 billion yen (~US$2.4 billion).[20] The impact of the floods, along with the impact of a September earthquake in Hokkaido, have been linked to a ~1.2 percent decline in Japan's 2018 gross domestic product (GDP). Japan's 2017 GDP was US$4.872 trillion. One percent of US$4.872 trillion is *US$48.72 billion*. Japan's economic slowdown, in turn, has combined with economic reductions in Germany and China to contribute to a potential global recession in 2019. Extreme precipitation is having substantial impacts on national and global economies.

Storyline-Based Attribution Studies: An Example for Houston

For the past several years, I have chaired sessions of talks focused on climate extremes at the annual meeting of the

[20] asia.nikkei.com/Politics/Japan-faces-2bn-price-tag-for-flood-rebuilding.

American Geophysical Union. Together with my co-chairs, I have been honored to hear presentations from some of the world's most eminent climate scientists. In this section I draw on talks by Dr. Michael Wehner from 2018 and 2019, because these talks provide excellent examples of extreme event attribution focusing on very harmful events. In general, this work tends to follow the "storyline" approach to attribution.[21] The storyline approach does not focus on questions like "will climate change make hurricanes more frequent," but instead begins with a specific event and asks the question "did climate change make this event more extreme."

Dr. Wehner's 2018 talk ("Causality and extreme event attribution – Or, was my house flooded because of climate change?")[22] began with a humorous homage to the "Rocky and Bullwinkle" cartoon, delved briefly into the epistemological foundations of climate attribution, and then presented some very powerful and important attribution analyses of hurricanes Harvey, Katrina, Maria, Irma, and Florence. Michael has kindly shared his presentation with me, and it is definitely worth our while to work our way through it – because it provides you with a front-row seat to the developing science of climate attribution.

Dr. Wehner began by noting that the causality of climate extremes is inherently complex. Extreme events are rare because multiple factors have to be aligned just right. In more ways than one, questions like "does smoking cause cancer" rhyme with "do CO_2 emissions enhance extreme climate events." Borrowing from epidemiology, climate scientists often ask one of two questions: (1) Has the probability of an event changed? (2) Has the magnitude of an extreme event changed? We answer these questions by considering two worlds: a world

---

[21] Trenberth, K. E., J. T. Fasullo, and T. G. Shepherd (2015). "Attribution of climate extreme events." *Nature Climate Change*, 5, 725–730. doi.org/10.1038/nclimate2657.

[22] The next several paragraphs follow very closely the material presented by Dr. Wehner.

with climate change and a world without industrialized humans. If we are interested in probability, we can fix the magnitude of an event and examine the change in the risk ratio of a given extreme. The risk ratio is one common measure of such a change: risk_ratio = [Probability in a world with climate change]/[Probability in a world without climate change]. For example, we might ask how climate change influences the probability that Houston would receive 50 inches of rain. Or, if we are interested in changes in magnitude, we can fix the probability and examine changes in quantity. We might ask, for example, how climate change alters the magnitude of precipitation in a one-in-twenty-year flood event.

Since we only have one planet, a planet *with* climate change, we have to rely on models to approximate a world without climate change. There are two basic approaches to doing this: climate models and statistical models. Attribution analyses using climate models typically follow an approach to causal theory developed by Judea Pearl, a professor at the University of California, Los Angeles. Just as in a medical trial, climate simulations are used to produce one group that gets a "drug" (climate change) and another "placebo" group that has boundary conditions representing a world without climate change. The first world is warmer and has increased greenhouse gasses. The second world is cooler and lacks human-induced increases in greenhouse gasses and aerosols. Attribution analyses can also be based on Granger Causality, named after Sir Clive Granger (1934–2009). These analyses are based on statistical inference, and are often used in medicine when explicit interventions are unethical. For example, feeding children high doses of sugar to see if they develop diabetes would be wrong. To explore such situations, we often use statistical models. In general, Pearl statements of causality are stronger than Granger statements, because we have carried out controlled experiments that can reduce the possible influence of hidden covariates.

Dr. Wehner presented examples of both types of causal analyses. Examining Hurricane Harvey (Figure 7.6), he reported

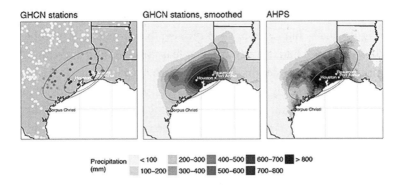

Figure 7.6 A reproduction of Figure 1 from Risser and Wehner's 2017 paper, "Attributable Human-Induced Changes in the Likelihood and Magnitude of the Observed Extreme Precipitation during Hurricane Harvey." Precipitation totals (mm) for the Houston, Texas, region from August 25 to 31, 2017 based on station observations (left), interpolated station observations (middle), and a radar station combination (right).

in a paper published in collaboration with Mark Risser.[23] This work used a statistical model (Granger attribution framework) to suggest that climate change increased the probability of such an extreme event by 350 percent (3.5 times). Examining changes in magnitude, Risser and Wehner found that climate change probably increased the magnitude of Harvey by something like 38 percent. In a world without climate change, Harvey might have brought 31 inches of rain, not 50.

Wehner next presented another compelling study, "Anthropogenic Influences on Major Tropical Cyclone Events," published by Christina Patricola and Michael Wehner in the journal *Nature*.[24] In this elegant "Pearl Causality" study, the scientists used a sophisticated high-resolution convection-resolving model to examine the behavior of hurricanes Katrina,

---

[23] Risser, Mark D., and Michael F. Wehner. "Attributable human-induced changes in the likelihood and magnitude of the observed extreme precipitation during Hurricane Harvey." *Geophysical Research Letters* (2017).

[24] Patricola, Christina M., and Michael F. Wehner. "Anthropogenic influences on major tropical cyclone events." *Nature* 563.7731 (2018): 339. www.nature.com/articles/s41586-018-0673-2

Irma, and Maria in worlds with and without climate change (Figure 7.7). Big science, big money, big impacts; remember these cyclones had costs totaling at least $446 *billion* and caused more than 5,000 deaths (Tables 7.1 and 7.2). While Patricola and Wehner's study didn't find big changes in wind speeds, they did find significant changes in precipitation totals. The left column of Figure 7.7 shows estimates of how much stronger the rainfall rates of Katrina, Irma, and Maria were because of human-induced warming of the oceans. The right column shows the change between the simulated hurricane intensities, based on the observed conditions, and simulated rainfall intensities at the end of the twenty-first century if current emission patterns continue.

As bad as Katrina, Irma, and Maria were, such cyclones are likely to get substantially worse if we do not curtail our emissions. Patricola and Wehner's study, in fact, found large (8–42 percent) increases in the end-of-the-twenty-first-century rainfall intensity in eleven cyclones (Katrina, Irma, Maria, Bob, Floyd, Gilbert, Ike, Iniki, Haiyan, Yasi, and Gafilo), assuming continued rapid greenhouse gas emissions. We are already seeing a rapid rise in the economic costs of hurricanes (Figure 7.3–7.5) as coastal development coincides with more intense rainfall rates. The impacts of flooding and intense rainfall, furthermore, can be nonlinear. Houston might be able to cope with a doubling of rainfall if a hurricane brings 24 inches of rain as opposed to 12 inches (61 cm versus 30.5 cm). The doubling from 24 to 48 inches (from 61 to 122 cm) brings with it much greater economic and societal destruction. What if future Harvey-type storms bring 75 inches (191 cm) of rain? Investing in reducing emissions *now*, while also making our energy economies more efficient, resilient, and effective, will likely be a smart investment.

Conclusion: Climate Change Is Hurting People Now

The Greek philosopher Solon (Figure 7.1) gave the city of Athens its famous set of laws, which proved so popular that

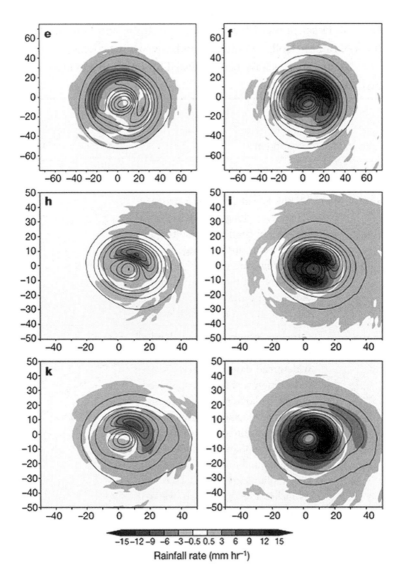

Figure 7.7 An extract of figure 3 from Patricola and Wehner's Nature paper. The rows correspond to Katrina, Irma, and Maria. The left-hand column shows the estimated change in rainfall rates in a world with and without climate change. The right-hand column shows results for the world as it was and the world as it will be if we continue our current pattern of emissions.

Table 7.1. Description of Harvey, Katrina, Irma, and Florence from the US Billion Dollar Disaster Database (www.ncdc.noaa.gov/billions/events). Losses have been adjusted for inflation and are shown 2018 US dollars.

Hurricane	Description	Economic Loss (billions)	Deaths
Katrina August 2005	Category 3 hurricane initially impacts the US as a category 1 near Miami, Florida, then as a strong Category 3 along the Louisiana–Mississippi coastlines, resulting in severe storm surge damage (maximum surge probably exceeded 30 feet) along the coasts, wind damage, and the failure of parts of the levee system in New Orleans.	$168.8	1,833
Harvey August 2017	Category 4 hurricane made landfall near Rockport, Texas, causing widespread damage. Harvey's devastation was most pronounced due to the large region of extreme rainfall producing historic flooding across Houston and surrounding areas. More than 30 inches of rain fell on 6.9 million people, while 1.25 million experienced over 45 inches and 11,000 had over 50 inches, based on 7-day rainfall totals ending August 31. This historic US rainfall caused massive flooding that displaced over 30,000 people and damaged or destroyed over 200,000 homes and businesses.	$130	89
Maria September 2017	Category 4 hurricane made landfall in southeast Puerto Rico after striking the US Virgin Island of St. Croix. Maria's high winds caused widespread devastation to Puerto	$93.7	2,981

Table 7.1. cont'd

Hurricane	Description	Economic Loss (billions)	Deaths
	Rico's transportation, agriculture, communication, and energy infrastructure. Extreme rainfall up to 37 inches caused widespread flooding and mudslides across the island. The interruption to commerce and standard living conditions will be felt for a long period as much of Puerto Rico's infrastructure is being rebuilt. Maria was one of the deadliest storms to impact the US, with numerous indirect deaths in the wake of the storm's devastation.		
Irma September 2017	Category 4 hurricane made landfall at Cudjoe Key, Florida, after devastating the US Virgin Islands – St John and St Thomas – as a category 5 storm. The Florida Keys were heavily impacted, as 25% of buildings were destroyed while 65% were significantly damaged. Severe wind and storm surge damage also occurred along the coasts of Florida and South Carolina. Jacksonville, Florida, and Charleston, South Carolina, received near-historic levels of storm surge causing significant coastal flooding. Irma maintained a maximum sustained wind of 185 mph for 37 hours, the longest in the satellite era. Irma also was a category 5 storm for longer than all other Atlantic hurricanes except Ivan in 2004.	$52	97
Total		$445.5	5,000

Table 7.2. Large cyclone disaster statistics from the Munich Re catastrophe database. Losses are shown in inflation-adjusted 2014 US dollars.

Cyclone	Date	Location	Economic Loss (millions)	Deaths
Enawo	March 2017	Madagascar	$200	81
Ava	January 2018	Madagascar	$130	51
Okchi	November 2017	India	$400	89
Gaja	November 2018	India	$780	57
Roanu	May 2016	Sri Lanka	$500	96
Vardah	November 2015	India	$3,500	597
Titli	October 2018	India	$920	85
Damrey	November 2017	Vietnam	$650	108
Doksuri	September 2017	Vietnam	$480	14
Mujigae	October 2015	China	$4,000	20
Hato	August 2017	China	$3,600	21
Meranti	September 2016	China	$3,000	29
Nepartak	July 2016	China	$1,400	83
Soudelor	August 2015	China	$2,2000	26
Chan-Hoim	July 2015	China	$1,4000	1
Lionrock	August 2016	South Korea	$100	138
Jebi	September 2018	Japan	$13,000	11
Trami	September 2019	Japan	$4,000	4
Pam	March 2015	Vanuatu	$300	11
Irma	September 2017	Cuba	$3,500	12
		Virgin Islands	$9,000	5
Mathew	October 2016	Cuba	$1,600	
		Haiti	$1,400	546
Maria	September 2017	Puerto Rico	$65,000	2,975
		Dominica	$1,200	31

Solon had to flee town or suffer reprisal. His many peregrinations brought him before wealthy Croesus, whom he told, "count no man lucky until his end is known." Not long after meeting Solon, Croesus was struck blind by the gods for hubris, his son was killed in a hunting accident, his empire was crushed by the Persians, and he ended his life on the business end of a funeral pyre while he was still breathing. We modern humans might very well be described as Croesian, but with the twist that our demise stems from our rise and will also primarily descend upon our descendants. Our tragedy follows the arc of Daedalus, the famed Greek craftsman whose wings fatally propelled his son Icarus too close to the Sun. As our human wealth explodes, we spew greenhouse gasses into our atmosphere, supporting an inevitable increase in both air temperatures and atmospheric water vapor. This increase in water vapor increases the intensity of extreme precipitation events. Hurricanes, cyclones, and typhoons are the most extreme and complex class of extreme rainfall events. While it is still difficult to detect changes in these events in the observational record, attribution studies such as those presented here for Harvey, Katrina, Irma, and Maria imply that we are already paying a steep economic and humanitarian cost for our greenhouse gas emissions. Table 7.1 shows the economic losses and fatalities from these five hurricanes. The attribution results presented here imply that a substantial fraction of these impacts resulted from human-induced warming. While detailed data at a global scale is more difficult to obtain, Table 7.2 shows similar statistics for large 2015–2018 catastrophes from the Munich Re database. Comprehensive attribution of these events is not available at present, but the balance of evidence suggests that there has been a detectable increase in the proportion of tropical cyclones reaching high levels of intensity, and anthropogenic forcing has contributed to this increase (Figure 7.8).

  Between absolute zero and absolute hot, our fragile Earth inhabits a small magic range of temperatures capable of supporting water in its frozen, liquid, and gaseous phases. Water transitioning from one phase to another supports the

Figure 7.8 Jacob Peter Gowy's The Flight of Icarus (1635–1637). en.wikipedia.org/wiki/Icarus#/media/File:Gowy-icaro-prado.jpg

development of tropical storms and cyclones. Cyclones collect water vapor from vast expanses and focus it into punishing rains, which in turn fuel intense winds and storm surge. The latest science suggests that a warming of a degree or two can lead to increases in tropical cyclone rainfall rates, intensity, and the proportion of storms that reach Category 4–5 intensity and enhance the severity of hurricanes like Harvey, Katrina, Maria, and Irma. Continued warming to the end of the twenty-first century may lead to further increases in the precipitation intensities of very strong cyclones. The delayed impact of our actions complicates Solon's definition of a happy life. We may die happy, but what of those who come after? Can we eat our vegetables, be kind, pleasant, and respected, and still die happy knowing that our actions set the stage for more catastrophic weather for Earth's future descendants?

8 CONCEPTUAL MODELS OF CLIMATE CHANGE AND PREDICTION, AND HOW THEY RELATE TO FLOODS AND FIRES

Introduction

Late 2019 brought catastrophic floods to East Africa. October to December (OND) 2019 rainfall totals were among the highest of the past nearly four decades in many areas in East Africa. Flooding and other related disasters impacted 3.4 million people in the region.[1] Nearly a million people were impacted in South Sudan, a country where extreme food insecurity is impacting almost the entire nation. In five other countries (Ethiopia, Somalia, Kenya, Uganda, and Djibouti), between 250,000 and 570,000 people were impacted. In early 2020, fires raged across Australia. As of January 4, 2020, hundreds of wildfires in Australia had scorched more than 12 million acres[2]; in New South Wales, the epicenter of the crisis, more than 480 million animals are estimated to have perished,[3] according to Sydney University ecologist Charles Dickman. Australia's

---

[1] OCHA January 2020 report, reliefweb.int/report/south-sudan/eastern-africa-region-regional-floods-and-locust-outbreak-snapshot-january-2020.

[2] www.usatoday.com/story/news/world/2020/01/03/australia-fires-map-animals-evacuations/2803057001/.

[3] sydney.edu.au/news-opinion/news/2020/01/03/a-statement-about-the-480-million-animals-killed-in-nsw-bushfire.html.

environment minister estimates that 30 percent (more than 8,400) of the koalas living in New South Wales' mid-north coast may have perished.

Connecting these floods and fires is a very strong positive Indian Ocean Dipole event. These events occur when the western Indian Ocean is very warm and the eastern Indian Ocean is very cool.

Understanding how climate change contributes to exceptional sea surface conditions will enhance our ability to understand, anticipate, and predict climate extremes like droughts and floods.

Stories matter. Stories shape our perception of the world. Stories mold our interpretation of the past. They inform our predictions of the future. In climate attribution studies, "storyline" approaches are being used to describe precipitation extremes.[4] Climate change increases the amount of water vapor in the atmosphere, which may enhance the severity of natural phenomena such as hurricanes.[5] As the atmosphere warms, we can expect more moisture to converge during many of the most extreme events.

We can consider extreme sea surface temperatures in a similar way, and such considerations can guide climate hazard predictions. Heat in the ocean, like water vapor in the atmosphere, is transported by currents. And like water vapor in the air, heat in the ocean can converge and dissipate due to circulation changes. But unlike the atmosphere, conditions in the world's oceans vary slowly. They persist. This persistence provides opportunities for prediction. Slow variations of sea surface temperatures will influence wind and water vapor patterns, offering opportunities for forecasts. When climate change contributes to extremely warm ocean conditions, we have windows

---

[4] Lloyd, Elisabeth A., and Naomi Oreskes. "Climate change attribution: When is it appropriate to accept new methods?" *Earth's Future* 6.3 (2018): 311–325. doi.org/10.1002/2017EF000665.

[5] Trenberth, K. E., J. T., Fasullo, and T. G. Shepherd. "Attribution of climate extreme events." *Nature Climate Change* 5 (2015): 725–730. doi.org/10.1038/nclimate2657.

of opportunity for early warning. To make the most of these opportunities, we need a clear understanding of how energy moves through the Earth's climate system. Understanding how climate change is actually altering the spatial distributions of energy and water vapor, week to week and month to month, can literally help us save lives.

Contrasting "Bathtub" Warming with Energy Convergence Patterns

Most of the "extra" energy provided by climate change goes into the oceans, and this extra energy is accumulating very rapidly.[6] The energy stored in the ocean accounts for about 93 percent of the total increase in the climate system since 1971. Figure 8.1 shows a time series describing the total amount of energy in the top 700 meters of the world ocean, expressed as anomalies (differences from the long-term mean). I look at this kind of data all the time, but I am yet again stunned by the rapid rate of change we are seeing in the top layers of the oceans. It took us about thirty years to go from an anomaly of about -3×10^{22} Joules in the 1960s to about $+3 \times 10^{22}$ Joules in the late 1990s. The current anomaly in 2019 is more than *five* times this amount. Between 2014 and 2019, we see another 5×10^{22} Joule jump. A 5×10^{22} Joule increase in energy is equivalent to the *energy released by about 12 million one-megaton nuclear bombs.*

To understand and anticipate climate extremes, it is critical to understand how this "extra" heat behaves. Energy moves around the oceans in complex ways. These movements can cause energy to concentrate in some places and dissipate in others. Such anomalies can produce climate hazards – persistent seasonal climate anomalies that put people and ecosystems in harm's way. But when the ocean is exceptionally warm in one place, it also tends to be cool somewhere else. Climate change will not remove the natural processes that lead heat in the oceans and

---

[6] www.climate.gov/news-features/understanding-climate/climate-change-ocean-heat-content.

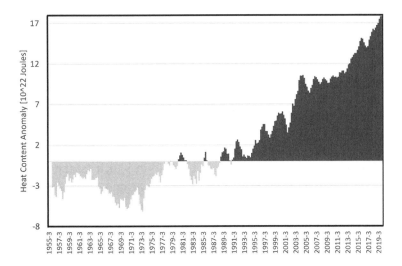

Figure 8.1 Global 0–700 m ocean heat content anomalies, based on differences from the long-term average global ocean heat content (1955–2006) in the top 700 meters of the ocean.[7]

water vapor in the atmosphere to accumulate in one region one day and another region the next. Importantly for early warning applications, the energy in the ocean changes very slowly because it takes such a huge amount of energy to heat water. This produces sea surface temperature gradients that persist at weekly and monthly time scales. These persistent gradients create opportunities for climate predictions and effective life-saving interventions.

Focusing on a conception of climate change as the average of collections of climate simulations can cause us to miss these opportunities.

Climate scientists often talk about "internal" and "external" sources of variability. "External" variations are represented by averages taken across a large number of climate simulations. These are variations that can be easily attributed to changes in greenhouse gasses, aerosols, land cover, and solar

---

[7] Global 0–700 m global ocean heat content anomalies, based on differences from the long-term average global ocean heat content (1955–2006) in the top 700 m of the ocean. data.nodc.noaa.gov/woa/DATA_ANALYSIS/3M_HEAT_CONTENT/DATA/basin/3month/ohc_levitus_climdash_seasonal.csv.

radiation. "Internal" variations are then identified as the deviations between the individual simulations and the ensemble. While this decomposition is useful, especially when applied to very broad area estimates, like the average temperature of the globe, or low-frequency changes over decades, it can rapidly lose its utility as we move to smaller and smaller spatial and temporal scales. Here, we will refer the spatially similar warming pattern produced by averaging many simulations as "bathtub" warming. Imagine a bathtub in which all the water warms at a uniform rate. While such a warming is a common way to think about climate change, we need to recognize the limitation inherent in this conception.

Global warming arises through the combined action of an incredible number of individual photons. Each ray of light that passes through a column of atmosphere with more greenhouse gasses leaves a little more heat behind. But the oceans absorb, assimilate, and translate this heat energy from place to place. So, if we are interested in understanding climate extremes, then the "bathtub" warming signal that we get from averaging across all of our climate change simulations, across all the surface of the Earth, provides a misleading "search pattern" – misleading in the sense that it is unlike each of the individual simulations that were actually produced by our climate change model – and misleading in the sense that there is no physical mechanism in the models or in the real world that accounts for an even flat "bathtub" warming signal across all the world's oceans. This "bathtub" pattern arises from averaging many individual simulations, but it does not actually exist from the perspective of actual physics.

According to the bathtub warming paradigm, sea surface temperature gradients will be similar to those in the past, since everywhere in the ocean will be warming at about the same rate, at the same time. If, on the other hand, human-induced warming increases the ocean heat transports associated with natural climate variations, enhancing the temperatures of naturally occurring, very warm sea surface temperatures, then what we will likely see are more pockets of very warm sea surface temperatures, and associated severe climate extremes – which

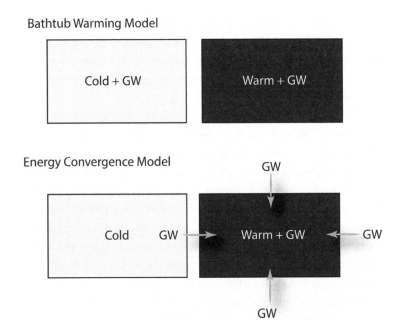

Figure 8.2 Two conceptual models of climate change. The top schema depicts a "bathtub" warming scenario where two ocean areas (one warm and one cold) each warm at the same rate. The bottom schema depicts an energy convergence model in which "extra" heat energy converges into a warm region of the ocean.

will provide opportunities for prediction. This is why philosophy matters. We need a conceptual model that assumes that heat energy will build up in the oceans and move around. This is increasing the frequency of extremely warm sea surface temperatures, which can induce climate extremes, droughts, and floods, giving us an opportunity for early warning. We need to be on the lookout for areas of extreme tropical or subtropical sea surface temperatures that lie next to naturally occurring regions of anomalously cool waters.

Figure 8.2 describes these competing conceptual models of climate change. In the top schema, an evenhanded "bathtub" warming affects a cold and warm region in the tropical and subtropical ocean. Please note that in this conceptual model, I am really referring to anomalously warm or cold regions of the tropical and subtropical oceans, not the difference between polar regions and the tropics. In the tropics, changes in sea surface temperatures are closely linked to changes in atmospheric circulation, and associated variations in rainfall and air temperatures.

Under a bathtub interpretation, the greenhouse gas warming (GW) would be about the same in both regions, leaving the sea surface temperature gradient between the two regions unchanged. In the bottom schema (the Energy Convergence Model), extra energy from greenhouse gas warming is transported by natural current variations into the warm region. This can lead to stronger temperature gradients. Temperatures in the cold region might resemble those in a world with less global warming, and those in the warm region could be much warmer than we might expect, given the average of many climate change simulations.

Thought experiments can help us understand why such conceptual models matter.

Let's set the mood. Smooth Jazz plays in the background. You don a swimsuit and slide into the experimental hot tub. Bath salts make the water pleasantly buoyant, and you float contentedly on your back.

> *Experiment* 1.
> *The bathtub warms slowly. You relax, adjusting to the heat, and it is 2080 before you realize the water is scalding hot. You are now a boiled frog.*
>
> *Experiment* 2.
> *You rest contentedly in a bathtub, gently dozing off toward sleep, the soulful burr of sappy Jazz saxophone pulling you slowly toward somnolence, when suddenly after 6,000 soapy soothing seconds — ow! Burning hot waters burn your feet. Then ow! Burning hot waters burn your bum. Then ow! Burning hot waters excoriate your hand. It is 2020 and you recognize that human-induced increases in hotness are hurting people. Now. You act.*

Experiment 2 describes one important manifestation of climate change. Rather than a bathtub warming, we are experiencing limited areas of exceptionally warm water that move from place to place to place, from month to month, lying alongside ocean waters with relatively cool "normal" temperatures.

Despite this (hopefully) humorous description, this topic is absolutely deadly serious. In the tropics, since 2015, pockets of exceptionally warm ocean waters have destroyed tropical coral reefs[8] and contributed to severe droughts in Eastern and Southern Africa.[9] On December 5, 2016, such an interpretation of climate change, together with a decade of research and cross-validated statistical models, was used to predict a high probability of back-to-back droughts in East Africa.[10] This work contributed to an effective and early humanitarian response, helping prevent a repeat of the 2011 Somali famine.[11] Here, however, we examine October–November 2019 conditions, and how they related to flooding and extreme air temperatures in East Africa and Australia.

Exceptional Flooding and Temperatures in 2019

In late 2019 (October and November), East Africa experienced severe flooding while Australia experienced extreme air temperatures and drought, conditions associated[12] with the Indian Ocean Dipole (IOD),[13] the gradient in tropical

---

[8] journals.ametsoc.org/doi/pdf/10.1175/BAMS-ExplainingExtremeEvents2016.1, Chapters 2, 9, 28.
[9] Funk, Chris, et al. "Examining the role of unusually warm Indo-Pacific sea-surface temperatures in recent African droughts." *Quarterly Journal of the Royal Meteorological Society* 144 (2018): 360–383. rmets.onlinelibrary.wiley.com/doi/full/10.1002/qj.3266
[10] Funk C., G. Husak, D. Korecha, G. Galu, and S. Shukla. (2016). Below normal forecast for the 2017 East African long rains, December 5, 2016. blog.chg.ucsb.edu/?m=201612.
[11] Funk, Chris, et al. "Recognizing the famine early warning systems network: over 30 years of drought early warning science advances and partnerships promoting global food security." *Bulletin of the American Meteorological Society* 100.6 (2019): 1011–1027.journals.ametsoc.org/doi/pdf/10.1175/BAMS-D-17-0233.1.
[12] www.bbc.com/news/science-environment-50602971.
[13] www.bom.gov.au/climate/iod/.

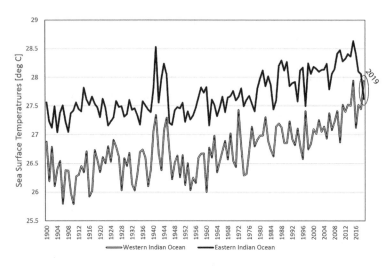

Figure 8.3 Time series of western and eastern Indian Ocean sea surface temperatures. Based on the NOAA Extended Reconstruction version 5 data set. The selected western Indian Ocean region stretched from 40°E to 80°E and 20°S to 3°N. The eastern Indian Ocean region stretched from 80°E to 110°E and 15°S to 3°N.

sea surface temperatures between the western and eastern Indian Ocean.[14] When the western Indian Ocean is extremely warm and the eastern Indian Ocean is extremely cold, eastern Africa receives torrential rains, while Australia experiences drought. In Figure 8.3, I display time series of sea surface temperatures from the tropical western Indian Ocean (20°S–3°N, 40°E −80°E) and the tropical eastern Indian Ocean (15°S–3°N, 80°E −110°E) during October–November. These boxes are a little different than the boxes typically used to define the typical IOD index, and were selected to correspond with the 2019 anomaly pattern, which featured an exceptionally strong Indian Ocean sea surface temperature gradient just south of the equator.

---

[14] Saji N. Hameed. (2018). The Indian Ocean Dipole, Oxford Research Encyclopedias, February 2018. DOI: 10.1093/acrefore/ 9780190228620.013.619. oxfordre.com/climatescience/view/10.1093/acrefore/ 9780190228620.001.0001/acrefore-9780190228620-e-619.

As you can see from Figure 8.3, historically, the eastern Indian Ocean has been substantially warmer than the western Indian Ocean. In the tropical oceans, regions with warmer sea surface temperatures tend to be much rainier than regions with relatively cool sea surface temperatures. So, under normal conditions, the eastern Indian Ocean and countries like Indonesia receive copious amounts of rainfall. The western Indian Ocean and countries like Somalia and Kenya receive relatively little rain.

But Figure 8.3 indicates a very large deviation from normal conditions in 2019. According to this data set, *for the first time since 1900, the western Indian Ocean was warmer, in absolute magnitude, than the eastern Indian Ocean.* The exceptionally warm western Indian Ocean sea surface temperatures occurred alongside atypically cool eastern Indian Ocean waters. We can observe an upward trend in the time series from both regions. Given this trend, and the natural historical variability, the exceptionally warm western Indian Ocean conditions in 2019 (and 2015) are not unexpected, but still exceptionally warm. The 2015 and 2019 values were the warmest on record, and much warmer (~+0.5°C) than any previous values.

But the behavior of the eastern Indian Ocean time series is also notable and important. It indicates that despite global warming, eastern Indian Ocean temperatures can still be quite cold – in this case, attaining values similar to those experienced in the 1960s. These results align with the Energy Convergence Model (Figure 8.2). We should expect pockets of exceptional warmth accompanied by adjacent cool areas. While both the western and eastern Indian Ocean are warming, there is no magical warming process that is heating both ocean regions by the same amount on a week-to-week or month-to-month basis.

Taken together, the combination of exceptionally warm western Indian Ocean sea surface temperatures and atypically cool eastern Indian Ocean *inverted* the Indian Ocean temperature gradient (Figure 8.3), producing a strong east-west pressure gradient that drove winds and water vapor westward across the equatorial Indian Ocean, feeding floods in East Africa but

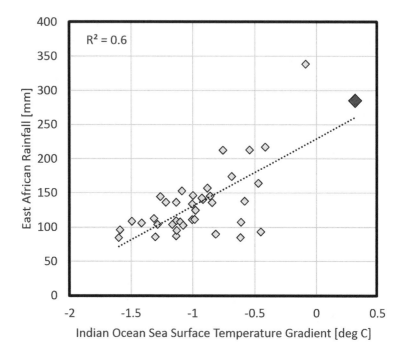

Figure 8.4 Scatterplot showing the relationship between 1981–2019 October–November Climate Hazards center Infrared Precipitation with Stations (CHIRPS) data from Uganda, Kenya, Somalia and Ethiopia south of 11°N and the east-west gradient of tropical Indian Ocean sea surface temperatures.

reducing rainfall and increasing air temperatures over Australia. Australia experienced the driest September-to-November spring rains on record (since 1900), according to the Bureau of Meteorology.[15]

Historically, there has been a strong relationship between October and November East African rainfall and the gradient between the western and eastern Indian Ocean sea surface temperatures (Figure 8.4). The strength of this relationship is primarily determined by behavior at the extremes. On the left of this scatterplot, we find a cluster of strong negative

---

[15] www.abc.net.au/news/2019-12-02/bureau-of-meteorology-declares-spring-2019-the-driest-on-record/11755848.

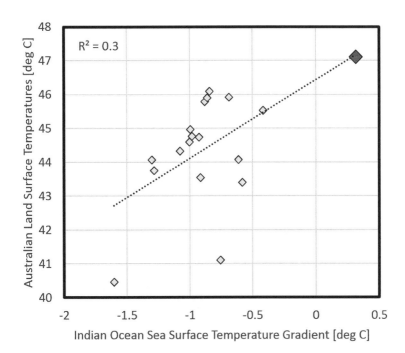

Figure 8.5 Scatterplot showing the relationship between 2002 and 2019 October–November Australian MODIS Land Surface Temperature data and the east-west gradient of tropical Indian Ocean sea surface temperatures.

gradient values and very low East African rainfall values. These years (2010, 1996, 1998, 1990, 1995, 2005, 2016) correspond with very dry East African rainy seasons. On the right of Figure 8.4, we find 1997 and 2019, two exceptionally wet years associated with extremely weak Indian Ocean sea surface temperature gradients. Figure 8.5 shows a similar but weaker relationship between satellite observations of Australian Land Surface temperatures.

Appreciating how exceptional these climate conditions were might have enhanced disaster preparedness and response. The stories we tell ourselves matter. While accurate numerical predictions are always an important aspect of effective early warning, the "holy moly" dimension is important as well ("holy moly" = an exclamation to express surprise or astonishment).

When we recognize that the tropical oceans are in an exceptional state, we are more likely to be on the lookout for potentially related weather and climate extremes. This is why identifying exceptional sea surface states can be so important. Consider the following illustrative statement that *could* have been made on November 5.

> October sea surface temperatures in the western Indian Ocean were exceptionally warm (27.5°C), the warmest on record, tying the previous extreme experienced in 2015. At the same time, October sea surface temperatures in the eastern Indian Ocean were extremely cool, on par with conditions experienced during the very strong Indian Ocean Dipole/La Niña event in 1997. Taken together, these anomalies produced an extremely strong Indian Ocean Dipole event, historically associated with flooding in East Africa and drought in Australia.
>
> October rainfall totals in East Africa were exceptionally wet, 162 millimeters, more than twice the 1981–2010 average and similar to the extremely wet 1997 rainy season. In Australia, October Land Surface temperatures were the warmest on record. They were also extremely warm in an absolute sense (45.8°C or 114.4°F), +2.9°C warmer than the 2000–2018 mean. Please note that these are estimates of the emission temperature of the land surface, not 2-meter air temperatures, which are typically a little cooler. Still, the fact that an entire continent had a land surface temperature of 45.8°C is truly exceptional and concerning. Historically, for both East African rainfall and Australian land surface temperatures, regional time series of October and November conditions are very well correlated. Correlations of about 0.8 indicate very strong levels of persistence from one month to the next. October Indian Ocean gradient values also exhibit reasonably high levels of correlation with November East

African rainfall and Australian land surface temperatures (correlations of R = 0.7 and 0.5). These relationships, combined with an appreciation of the exceptional October–November ocean state, support predictions of exceptionally wet East African and warm Australian conditions in November 2019. Figures 8.6 and 8.7 show the regression-based forecasts supporting these statements.

These relatively simple results illustrate how monitoring extreme rainfall, land surface temperatures, and sea surface temperatures can inform skillful one month outlooks of potential weather extremes.

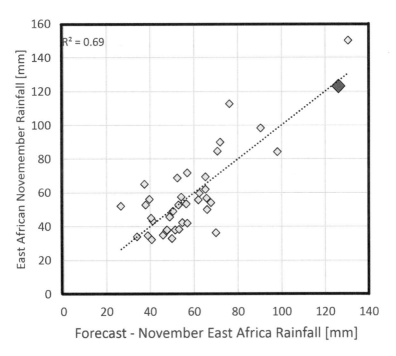

Figure 8.6 Scatterplot showing predicted and observed 1981–2019 November East African precipitation. Based on Climate Hazards center Infrared Precipitation with Stations (CHIRPS) data from Uganda, Kenya, Somalia, and Ethiopia south of 11°N, October CHIRPS data and the October east-west gradient of tropical Indian Ocean sea surface temperatures.

178 / Drought, Flood, Fire

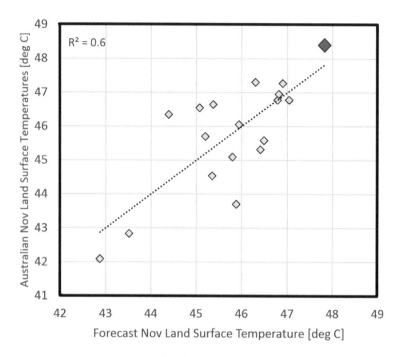

Figure 8.7 Scatterplot showing predicted and observed 2002–2019 November Australian MODIS Land Surface Temperature values along with October forecasts based on the east-west gradient of tropical Indian Ocean sea surface temperatures and October Australian land surface temperatures.

Climate Change?

Was the October–November 2019 Indian Ocean gradient event caused by climate change? Almost certainly not. Historically, we see that there have been strong east-to-west gradient variations for as far back as the data goes. But did climate change increase the *strength* of the 2019 gradient event? I would say almost certainly yes. Recent studies have used climate change simulations to examine the frequency of strong positive (warm western Indian Ocean) Indian Ocean Dipole events in a warming climate. Even under a very modest 1.5°C

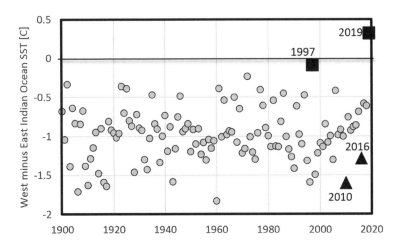

Figure 8.8 Time series of the gradient between October and November western and eastern Indian Ocean sea surface temperatures. Signature flat gradient years associated with flooding in East Africa are shown with squares. Strong negative gradient years associated with East African droughts are identified with triangles. The year 2019 is the only one since 1900 in which the west was warmer than the east.

warming scenario, these models indicate a doubling in the number of extreme IOD events.[16]

Looking at the observed 1900–2019 sea surface temperatures (Figure 8.3) we can see that the 2019 and 2015 western Indian Ocean temperatures were truly exceptionally, about 0.5° warmer than even the warmest prior values. We can see a strong tendency toward warmer conditions in the western Indian Ocean, which is one of the fastest warming ocean regions, with more warming projected by models.

These historically unprecedented warm conditions combined with "naturally" cool eastern Indian Ocean conditions combined to produce a stunning reversal of the equatorial Indian Ocean sea surface temperature gradient (Figure 8.8).

---

[16] Cai, Wenju, et al. "Stabilised frequency of extreme positive Indian Ocean Dipole under 1.5 C warming." *Nature Communications* 9.1 (2018): 1419.www.nature.com/articles/s41467-018-03789-6.

Typically, the eastern Indian Ocean is warmer than the western region (Figure 8.3), so this gradient has a 1900–2019 mean value of −1°C. Recent La Niña seasons like 2010 and 2016 were associated with dry October–November conditions in East Africa. Before 2019, only one year (1997) was associated with a very weak (near zero) gradient. According to this data set, 2019 is the only year since 1900 in which the western Indian Ocean was actually warmer than the eastern Indian Ocean. The climate change–enabled exceptional warming in the west (Figure 8.3) combined with naturally occurring cool conditions in the east to produce a truly historic reversal of the Indian Ocean sea surface temperature gradient and extreme East African precipitation (Figure 8.6).

Conceptual Models Can Lead to Rapid Transformation

How we think about climate change influences how we see the world. If we are unduly wed to a fixed concept of climate change as a bathtub-like warming of the land and oceans, then we may miss the opportunities for prediction provided by exceptional warming events. Western Indian Ocean temperatures (Figure 8.3) and the associated inversion of the Indian Ocean sea surface temperature gradient (Figure 8.8) provide a compelling example.

There are times when averaging can obscure the truth. Climate change involves much more than just the average trends produced by large collections of climate change model simulations. Extreme temperature gradients, such as those shown in Figure 8.8, might arise within these simulations, and are anticipated by climate change simulations. Averaging across simulations would obscure these patterns.

Consider Figure 8.9, which shows images of individual snowflakes. Each flake represents a unique and beautiful manifestation of complex processes.

181 / Conceptual Models of Climate Change and Prediction

Figure 8.9 Pictures of individual snowflakes – each unique.

Now consider Figure 8.10, which shows the average of all twelve snowflakes. In a very real way, this image is wrong. It creates an image of a snowflake that never really existed. Averaging men and women to find a "typical" human also does not make sense. We would not average all the animals in a zoo to create a "standard animal" and use that average as our expectation for what the next animal we meet is going to look like. In the same way, averaging across climate simulations can

Figure 8.10 An average of twelve snowflakes. Something that never happens in nature.

obscure important extreme climate states, such as stronger El Niños, La Niñas, and positive Indian Ocean Dipole events. Weather events, like snowflakes, arise through the complex nonlinear interplay of fluxes and forces. Averaging across these patterns does not necessarily represent a "true" manifestation of nature. While taking averages over large numbers of climate change simulations may provide insights into the low-frequency tendencies of sea surface temperatures, these averages may also obscure the important behavior of the extremes.

For effective early warning, it is important to remember that there is no "external forcing" that acts to ubiquitously heat the ocean surface in a consistent manner. The actual greenhouse longwave radiation heating at the ocean's surface is quite small (~2–3 Watts per meter squared), so it is the net accumulation of heat, and its tendency to concentrate in one place or another that tends to drive many twenty-first-century climate hazards.

Sharpening our ability to recognize extreme ocean temperatures, and their predictable influence on climate, will enhance our capacity to mitigate disasters. Such recognition can also reveal the tragic moral consequences of our greenhouse gas emissions, but may also lead toward a sustainable future.

We are what we pay attention to. This focus can alter quickly, triggering rapid societal change. For example, during the Renaissance, science grew up alongside the humanities and advances in technology. Scholars and thinkers began looking at the world as it was, not just the world as it was written. Art and architecture helped trigger scientific revolution. Inspired by Humanist thinkers and the art of ancient Greece and Rome, fifteenth-century artists like Brunelleschi and Alberti invented linear perspective, achieving convincing representations of depth. Leonardo Da Vinci observed the old world through new eyes.

Strolling through a museum, like flicking on a light, we can see the Renaissance's rapid transition in how painters began to see and render the human form. What was flat before became full of life. Marvelous architecture and universities sprung up in cities such as Florence, Bologna, Venice, and Milan. At the University of Pisa, Galileo Galilei and his pupils studied "disegno" (drawing) and later taught perspective and chiaroscuro, the new Renaissance technique of using strong contrasts in lighting to create striking representations of three-dimensional forms. Galileo also pursued an avid interest in mathematics, and in 1592 became a professor of geometry, mechanics, and astronomy at the University of Padua.

Like many Renaissance scientists, Galileo sought to uncover God's mysteries as written in the "book of the world," a source of truth as important as the Bible. Unique to Galileo, however, was the newfound desire to express that knowledge in the new language of mathematics. Or, to quote Galileo's *Assayer*,

> Philosophy [nature] is written in that great book which ever is before our eyes – I mean the universe – but we cannot understand it if we do not first learn the language and grasp the symbols in which it is written. The book is written in mathematical language, and the symbols are triangles, circles and other geometrical figures, without whose help it is impossible to comprehend a single word of it; without which one wanders in vain through a dark labyrinth.

184 / Drought, Flood, Fire

Figure 8.11 Fra Angelico's and Sandro Botticelli's depiction of the Madonna circa 1440 and 1481.

Once attuned to this new tome, humanity advanced quickly. Isaac Newton (1642–1726) literally brought the heavens down to Earth, showing that the same force of gravity that worked on Earth could describe and predict the motions of the planets.

Modernity, science, and progress ensued, liberating millions and then billions from aching toil and mind-numbing repetition. But now we face the painful impacts of our own success as climate change contributes to more extreme climate anomalies. But together we can face this challenge by getting better at seeing the impacts of climate change, and getting better at seeing the world around us.

History suggests we can do this. Consider, for example, Fra Angelico's depiction of the Madonna from circa 1440 and Sandro Botticelli's painting from 1481 (Figure 8.11). Like flicking on a light, we can see the rapid transition in how painters began to see and render the human form. In just 40 years, the way artists perceived and presented the human form blossomed. A similar focus on improving our perception and prediction of climate extremes will help us survive the rest of the twenty-first century. Keeping an eye out for areas of exceptionally warm sea surface temperatures will help us provide effective predictions. The "fingerprints" of climate change already form a heavy hand. But climate-smart early warning systems may help us mitigate some of the worst impacts.

9 CLIMATE CHANGE MADE THE 2015–2016 EL NIÑO MORE EXTREME

The story of this book began with a dream.

August 1995, Chicago, 2 AM. Air so wet and warm it blends undistinguished from the waters of Lake Michigan, which lap against my thighs. Facing the fullest moon, the city lights fan behind me north, south, and west. Behind me waves lap gently at the shore, and late-night cars roll along Shoreline Drive. Laughing and drinking cheap American beer, we are young, awake, and partying while others snooze away the night in fetid dampness. Perfect moment. But then begins the strangest pull. Gently against my legs comes a soft insistent tug, a soft push against the back of my things. The push grows stronger. My first thought: tide must be going out. Tug turns to tow, intensifies. My second thought: Oh crap... Lake Michigan is not tidal.

I could have raced to a pay phone. I should have flagged down a policeman.

Instead I leapt aboard my motorcycle, fired it up, and raced to my office in Chicago's south loop financial district.

Land-locked Lake Michigan is not tidal. The rapidly receding waters foretold an imminent tsunami, a catastrophic flood. From my office on the ninth floor I phoned in my stock options, shorting the stock market. Sell sell sell, as disaster for my fellow Chicagoans loomed. Man, was I going to be rich!

Upon waking, this dream lingered. Like a limping shadow twin it followed me, haunting. Who had I become?

By the next fall I was leaving my job as a programmer. Leaving my great group of friends, my rock band, and all my stuff. With a hammock, sleeping bag, and cook pot strapped to the back of my little Honda motorcycle, I headed west, like so many before me.

Riding without a windscreen, I laughed into the wind of America's blue highways, spitting back the mosquitoes. Heading south across Missouri, every day's advance matched the Sun's autumnal retreat, so I seemed to race frozen in time, each day bringing the same turning of the leaves from green to gold. Fall itself seemed timeless, I moving with it, a southward amber traveling wave, frost following right behind.

By day leaning into the wind, by night camping under the stars, hammock slung between two trees, I came west the southern way. In north Texas I pulled over and helped some cowboy types kick out the makings of a lightning-strike forest fire beneath the pines. In northern New Mexico I stopped to see my dad. Delicious hikes high in the Sangré de Christo Mountains with the first snows deep along the rushing streams. Delicious pozole spiced with New Mexican chilies. Then came the beautiful drive across northern Arizona. Then in California, a three-day backcountry trek in mysterious Joshua Tree National Park, filled with twisted, blasted, barren beauty. Finally I arrived in Santa Barbara. I was going back to school. I was going to become a geographer. I was going to make maps of the future.

While I was studying at the University of California, Santa Barbara (UCSB), I met two of the most important people in my life: my wife (Sabina née Barattucci) and Jim Verdin. Sabina got me excited about living life. Jim got me excited about living work, about using satellites and climate information to save lives and livelihoods. Sabina and I married in 1999. I finished my PhD in 2002, and we brought into the world two wonderful creatures, my La Niña and El Niño (*Amelie and Thelonious*). I was working with Jim to support the Famine Early Warning Systems Network (FEWS NET, www.fews.net). FEWS NET provides unbiased, evidence-based analysis to governments and relief agencies who plan for and respond to humanitarian crises. The 2000–2001 southern African rainy season had been poor, and then in the fall of 2002 an El Niño developed. El Niño events occur when exceptionally warm sea surface temperatures form in the eastern equatorial Pacific. They typically produce droughts in southern Africa. At this time, FEWS NET did not make much use of forecast information. I had worked on this research as part of my PhD.

The time had come to refute my dark, options-buying, stock market–shorting shadow from that dream gone by. Already exhausted from twindom, I worked feverishly. Alongside Jim and my Zimbabwean friend and colleague, Tamuka Magadzire, I crafted a statistical forecast model and accompanying report, "Forecasts of 2002/2003 Southern Africa Maize Growing Conditions," which called for "the possibility of dry growing conditions in these regions, together with the particularly dry forecasts for Northeastern Republic of South Africa and Southern Mozambique, suggest an increased probability of a poor crop production season." Taking a red-eye flight to Washington, DC in December, I presented these results and helped motivate an effective humanitarian response in Zimbabwe, a country quickly eroding under the increasingly erratic guidance of Robert Mugabe.

Fast forward to September 2003, six months after the end of Southern Africa's 2002–2003 rainy season. Poor rains had contributed to widespread hunger and disruption in many poor countries like Zimbabwe. FEWS NET reported, "The 2002/03 harvest is running out for most rural households, and purchased foods are selling at prices that continue to escalate far beyond the reach of the majority of poor households."[1] Ironically, for many regions dependent on summer rains, the hungriest time of the year, or the lean period, arises when the next season's rains begin. The good news was that the World Food Programme (WFP) was already distributing food aid to almost two million Zimbabweans.

The timely distribution of such aid is a great human accomplishment, an example of humanity being our best selves.

---

[1] fews.net/sites/default/files/documents/reports/Zimbabwe.193"/>_200309en.pdf.

Reaching beyond barriers of race, religion, or nationality, we act effectively to save the lives and livelihoods of desperately hungry people. I thought, "This is way cooler than capitalizing on the imminent inundation of America's Windy City."

To appreciate why this relief matters, and why I am so deeply concerned about climate change intensifying droughts, you need to understand both how horrible severe hunger can be and how and why it remains so widespread. In disaster prevention and relief, precision enables action. We spend a lot of time quantifying how bad "bad" is. As a drought expert, one of my specialties is producing the data and analyses underlying statements like "Ethiopia's 2015 drought was the worst in 50 years"[2] or "Southern Africa's 2015/16 drought was the worst in 36 years."[3] Food security experts (which I am not) have also developed a means for quantifying extreme food insecurity. This "integrated phase classification" supports the comparison of food insecurity in diverse regions – supporting the comparison of conditions in very different countries. The classes[4] include: 1. Generally Food Secure; 2. Moderately/Borderline Food Insecure; 3. Acute Food and Livelihood Crisis; 4. Humanitarian Emergency; and 5. Famine. The determination of which class a household or community belongs to is based on a number of different criteria. This complexity is necessary, given the great diversity of our societies and the multiple pathways to extreme hunger. Under this classification scheme, famine is characterized by at least one in five households facing an extreme lack of food, more than 30 percent of children under five suffering from acute malnutrition (wasting), and at least two people out of every ten thousand dying each day from starvation and starvation-related health complications.

One simple metric of severe malnutrition can help us imagine the horrors of hunger. Young children are typically

---

[2] www.telegraph.co.uk/news/2016/04/23/ethiopia-struggles-with-worst-drought-for-50-years-leaving-18-mi/.

[3] www.unocha.org/story/el-ni%C3%B1o-southern-africa-faces-its-worst-drought-35-years.

[4] fews.net/IPC.

Figure 9.1 The width of an extremely undernourished child's forearm.

deemed to be suffering acute malnutrition when their upper arm circumference is less than 11.5 cm. That's an upper arm diameter of 3.7 centimeters, or 1.4 inches, about as big as this circle (Figure 9.1).

Think about the hunger, the horrible suffering, involved in a four- or five-year-old child with such a stick-thin forearm. Now multiply that pain by tens of thousands of people. It is pretty tough math. While there are multiple indicators linked to integrated phase classes, such as mortality rates, coping strategies, and food access, acute malnutrition in children under five using a measure like upper arm circumference is one of the easiest to measure and imagine. During a famine more than 30 percent of children under five may exhibit acute malnutrition. At the integrated phase class 4, 15 percent of children may exhibit acute malnutrition. At integrated phase class 3, 10–15 percent children may exhibit acute malnutrition.

Hunger, El Niño, and the Southern Oscillation

El Niños can induce droughts[5] in many regions, which can lead to famine. For example, El Niño–related famines in 1876–1878, 1897, and 1899–1902 struck India, China, Brazil,

---

[5] Glantz, M. H. *Currents of Change: Impacts of El Niño and La Niña on Climate and Society*. Cambridge University Press, 2001.

Ethiopia, Korea, Vietnam, the Philippines, and New Caledonia.[6] Some 19 million people in India and 10 million people in China may have perished due to these El Niño–related droughts. In 1899–1900, Indian harvests in the Bombay Deccan, Karnatak, and Gujarat provinces were only 4–16 percent of normal according to the Bombay government's "Report on the Famine in the Bombay Presidency."[7] Yet Victorian Britain continued to extract Indian cotton and wheat even as millions of children and adults wasted, withered, and died (Figure 9.2). Indian authorities, held under rigid, inflexible ideological British rule, failed to respond adequately to the extreme conditions. George Nathaniel Curzon, First Marquess Curzon of Kedleston, served as Viceroy. In his zeal to suppress Home Rule for India, Curzon tightened press censorship, clamped down on education, and pitted Hindu against Muslim. For Curzon, financing the Boer war in South Africa was much more important than relieving the distress of famine-stricken people of India. Writing at the time, and quoting data from the *Lancet*, William Digby wrote, "This statement by what is probably the foremost medical journal in the world means that the loss of life thus recorded represented the 'disappearance' of fully one-half a population as large as that of the United Kingdom."[6]

By 1899, India was supplying one-fifth of England's wheat, expanding cotton plantations, and financing Britain's Asian military interests as the Indian Army engaged in adventures in Ethiopia, Sudan, and Egypt. As noted by Nobel Prize–winning economist Amartya Sen, these crises were rooted in chronic poverty and unjust economies, not just poor crop production.[8] As Davis notes, "During the famine of 1899-1900, when 143,000 Berars died directly from starvation, the province exported not only thousands of bales of cotton but an

---

[6] Davis, Mike. *Late Victorian Holocausts: El Niño Famines and the Making of the Third World*. Verso, 2000.

[7] Report on the famine in the Bombay Presidency, 1899–1902: Vol. I – Report, dspace.gipe.ac.in/xmlui/handle/10973/38215.

[8] Sen, Amartya. *Poverty and Famines: An Essay on Entitlement and Deprivation*. Oxford and New York: Clarendon Press and Oxford University Press, 1982.

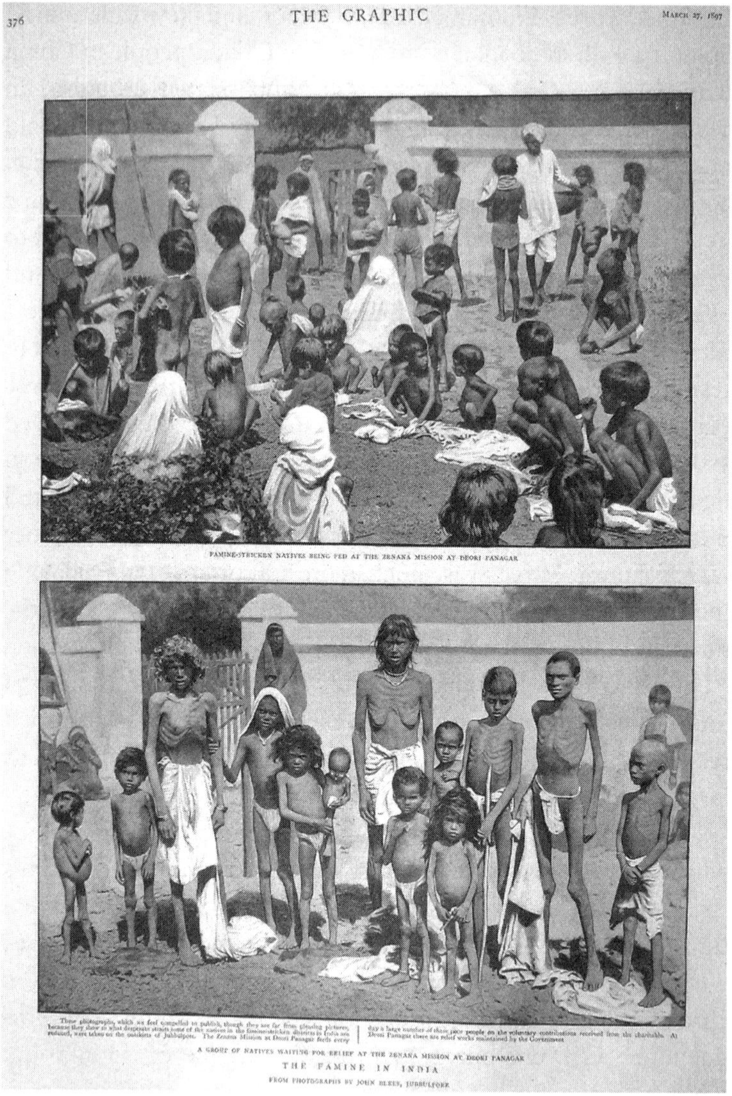
376 THE GRAPHIC MARCH 27, 1897

FAMINE-STRICKEN NATIVES BEING FED AT THE ZENANA MISSION AT DEORI PANAGAR

These photographs, which we feel compelled to publish, though they are far from pleasing pictures, because they show to what desperate straits some of the natives in the famine-stricken districts in India are reduced, were taken on the outskirts of Jubbulpore. The Zenana Mission at Deori Panagar feeds every day a large number of these poor people on the voluntary contributions received from the charitable. At Deori Panagar there are relief works maintained by the Government

A GROUP OF NATIVES WAITING FOR RELIEF AT THE ZENANA MISSION AT DEORI PANAGAR

THE FAMINE IN INDIA

FROM PHOTOGRAPHS BY JOHN BLEES, JUBBULPORE

Figure 9.2 Famine relief at the Zenana Mission in Deori Panager, on the outskirts of Jabalpur, India, March 1897.

incredible 747,000 bushels of grain." Today, El Niño events still pose large climatic risks, especially for relatively less well-off countries in the tropics. But our ability to understand, predict, and respond to extreme El Niños has increased substantially.

We will likely need all three dimensions of preparedness (understanding, prediction, and response) in the future, because climate change simulations anticipate more frequent and extreme El Niños.[9]

One huge step forward for modern climate science – and our ability to predict El Niño–related climate disasters – occurred in the latter half of the twentieth century, when scientists the likes of Jacob Bjerknes[10] connected the dots between the ocean phenomena of El Niños and the atmospheric variation known as the Southern Oscillation (SO). El Niños occur when the eastern equatorial Pacific becomes exceptionally warm. The term "El Niño" was coined by fisherman off the coasts of Peru and Chile during the seventeenth century. The appearance of unusually warm waters brought increased fish catches, and these beneficially warm waters were named "El Niños" or "Christ Child" because they typically occurred around the month of December. Much later, early in the twentieth century, the British meteorologist Sir Gilbert Walker coined the term "Southern Oscillation" to describe the inverse relationship between sea level pressures in Darwin Australia and Tahiti. Together, these two components describe the coupled ocean–atmosphere phenomena known as the "El Niño–Southern Oscillation"[11] or ENSO for short. Warm East Pacific El Niño waters occur alongside low atmospheric pressures near Tahiti, while on the other side of the ocean the West Pacific cools and air pressures near Darwin Australia increase. During a La Niña event, the opposite occurs. In the aftermath of the terrible turn-of-the-century Indian famines, Walker had been appointed as Director General of the Indian Observatories in 1903. A mathematician by training, Walker organized the Indian

---

[9] Cai, Wenju, et al. "Increasing frequency of extreme El Niño events due to greenhouse warming." *Nature Climate Change* 4.2 (2014): 111–116. www.nature.com/articles/nclimate2100.

[10] Bjerknes, J. "Atmospheric teleconnections from the equatorial Pacific." *Monthly Weather Review* 97 (1969): 163–172.

[11] Bjerknes, Jacob. "Atmospheric teleconnections from the equatorial Pacific." *Monthly Weather Review* 97.3 (1969): 163–172.

weather observatories and analyzed the accuracy of monsoon forecasts. Walker pioneered the use of lagged correlations as a means to make predictions. Such lagged correlations, as discussed in Chapter 8, remain important tools for early warning systems. Looking for predictive anomalies that could be used to forecast Indian droughts led Walker to analyze global weather variations.[12] This analysis led to his discovery of the Southern Oscillation or the "swaying of pressure on a big scale backwards and forwards between the Pacific Ocean and the Indian Ocean."[13]

Under normal conditions, as discussed in Chapter 2, the global Walker Circulation draws warm waters and moist air into the tropical regions surrounding Indonesia, producing the warmest ocean region on the planet. Winds blowing west across the eastern Pacific have a cooling effect, resulting in a strong gradient in average tropical sea surface temperatures, with the eastern Pacific being typically as much as five degrees cooler than the west.

But sometimes this pattern breaks down. The west-blowing winds weaken, west Pacific sea surface temperatures cool rapidly, while east Pacific sea surface temperatures increase dramatically. These changes are caused by, and cause, large-scale variations in the Indo-Pacific atmospheric circulation (Figure 9.3). Torrential rains follow, drawing air up and shifting the shape of wind patterns and storm systems around the world. Under normal conditions a strong oceanic temperature gradient (called a thermocline) stretches between Indonesia and the coast of Peru. Subsurface waters are much warmer near Indonesia. During an El Niño, a massive quantity of heat energy is shifted east, warming the sea surface, triggering a quasi-global disruption of the Earth's climate system.

---

[12] Walker, G. T. "Correlation in Seasonal Variations of Weather, VIII: A Preliminary Study of World Weather." Memoirs of the Indian Meteorological Department, 1923.

[13] Katz, R. W. "Sir Gilbert Walker and a connection between El Nino and statistics." *Statistical Science* 17 (2002), 97–112.

195 / Climate Change Made the 2015–2016 El Niño More Extreme

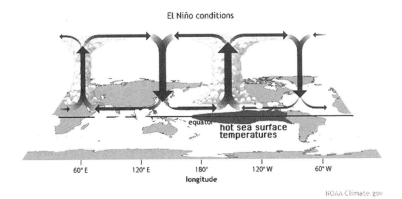

Figure 9.3 Generalized Walker Circulation (December–February) anomaly during El Niño events, overlaid on a map of average sea surface temperature anomalies. Anomalous ocean warming in the central and eastern Pacific (dark gray) helps shift a rising branch of the Walker Circulation to east of 180°, while sinking branches shift to over the Maritime continent and northern South America.
Source: NOAA Climate.gov drawing by Fiona Martin. From "The Walker Circulation: ENSO's atmospheric buddy," by Tom Di Liberto, www.climate.gov/news-features/blogs/enso/walker-circulation-ensos-atmospheric-buddy

El Niños are the biggest natural disruption in our climate systems, shifting weather in many parts of the world from their norm. To the west of the Pacific, El Niños can influence the Asian, African, and Australian monsoons, producing droughts in Indonesia, Thailand, India, Ethiopia, and Southern Africa. To the east of the Pacific, in Central America, the Caribbean, and northern South America, the chances of droughts are also increased. And almost everywhere, El Niños can cause air temperatures to jump upward as energy accumulated deep in the ocean is released.

All of these potential impacts unfurled during the 2015–2016 El Niño. Because of my job as an early warning analyst, I had a front row seat to much of this life-threatening mayhem. During the northern hemisphere summer of 2015 (June–August), life-sustaining monsoon rains typically sweep across the most populous regions of the globe, from northern Ethiopia to India to Thailand. During 2015, our Climate Hazards Center was watching Ethiopia closely. Since the early 2000s, I have been very personally invested in developing satellite and rain gauge–based data sets for monitoring droughts in Ethiopia. The El Niño–related 1983–1985 famine occurred when I was in high school, leading to more than a million deaths. Many of you may remember Live Aid. In 1999, I joined the FEWS NET science team, hoping to do my part to keep such a disaster from reoccurring. Being able to rapidly quantify how bad a drought is going to be is one of the key first steps in motivating an effective response. By combining satellite observations from space with long time series of rain gauge observations, we were able to determine that in the summer of 2015, northern Ethiopia had had its worst drought in fifty years.[14] FEWS NET reported on this crisis,[15] and a quick internet search of "Ethiopia's worst drought in 50 years" identifies articles on NPR, the *New York Times*, Catholic Relief Services, Voice of America, NBC, *Time Magazine*, CBC, UNICEF, Save the Children, and BBC.

As the summer harvest failed, food prices rose and the country plunged into one of the most severe food crises since 1984. In northeastern Ethiopia, plummeting livestock and milk production conspired with rising food prices to push millions to the edge of famine (Figure 9.4). Thankfully, massive international assistance began to arrive.[16] The government of Ethiopia assessed the required assistance as approximately US$1.4 billion. USAID alone contributed more than half a billion dollars in aid. The World Food Programme provided targeted supplementary feeding interventions to more than

---

[14] fews.net/sites/default/files/documents/reports/FEWS%20NET_WFP_Ethiopia.200"/>%20Alert_20151204.pdf.
[15] fews.net/sites/default/files/documents/reports/FEWS%20NET_Ethiopia%202015%20Drought%20Map%20Book_20151217_0.pdf.
[16] www.usaid.gov/sites/default/files/documents/1866/ethiopia_ce_fs07_03-30-2016.pdf.

Figure 9.4 Many animals have died as a result of the drought. Animal carcasses are common sight around the Fadeto and Hariso Internally Displaced Persons centers, Siti region, Ethiopia. The effects of a super El Niño are set to put the world's humanitarian system under an unprecedented level of strain in 2016 as it already struggles to cope with the fallout from conflicts in Syria, South Sudan, Yemen, and elsewhere. Oxfam estimates the El Niño weather system could leave tens of millions of people facing hunger, water shortages, and disease next year if early action isn't taken to prepare vulnerable people for its effects. It's already too late for some regions to avoid a major emergency. In Ethiopia, the government estimates that 10.2 million people will need humanitarian assistance in 2016, at a cost of $1.4 billion, due to a drought that's been exacerbated by El Niño. Courtesy of Abiy Getahun/Oxfam

450,000 children and pregnant and lactating mothers experiencing acute malnutrition. By March almost one million people required emergency water-trucking services to meet basic needs. More than half a million metric tons of wheat were transported to Ethiopia through the Port of Djibouti, and more than 11 million Ethiopians required relief food assistance.

In 2015, due to the monster El Niño, the Asian summer monsoon also fared poorly. In India, massive rainfall deficits stretched across much of the country, impacting 330 million people,[17] a quarter of India's population. A UNICEF report, "When coping crumbles – A Rapid Assessment of the Impact of Drought on Children and Women in India,"[18] focused on an analysis of 118 villages in drought-afflicted portions of nine states: Maharashta, Bihar, Madhya Pradesh, Chattisgarh, Telangana, Rajasthan, Jharkhand, and Odisha. This study found that "nearly 90 per cent of the people faced severe food shortage from January to June 2016." Farm incomes were destroyed, and most villagers relied solely on government subsidies from the Mahatma Gandhi National Rural Employment Guarantee Act.

In India and Pakistan an extreme (>45°C or >113°F) heat wave led to more than 2,500 deaths[19] (as discussed in Chapter 5).

Thailand experienced extremely warm, dry weather and crippling crop losses amounting to US$500 million in agricultural losses.[20] In July and October 2015, Indonesia experienced severe drying and extreme temperatures. Some 1.2 million people were classified as extremely food insecure, and a World Food Programme assessment found severe reductions in rice production.

Between June and December 2015, the El Niño grew stronger, and reached its peak in January 2016, contributing to yet another severe drought over Southern Africa – the worst drought there in thirty-six years. Poor October 2015 to March 2016 rains resulted in huge crop production deficits and large increases in food prices.[21] Extreme hunger threatened more than

---

[17] Guha-Sapir, D., et al. "Annual disaster statistical review 2016: The numbers and trends." Brussels, Belgium: Centre for Research on the Epidemiology of Disasters, 2016.
[18] reliefweb.int/sites/reliefweb.int/files/resources/pub_doc117.pdf.
[19] Wehner, Michael, et al. "S16. The Deadly Combination of Heat and Humidity in India and Pakistan in Summer 2015." *Bulletin of the American Meteorological Society* 97.12 (2016): S30–S32.
[20] Christidis, Nikolaos, et al. "The hot and dry April of 2016 in Thailand." *Bulletin of the American Meteorological Society* 99.1 (2018): S128–S132.
[21] www.sadc.int/files/3214/7806/7778/SADC_Regional_Situation_Update_No-3_Final_011116_V1.pdf.

24 million people. Northern South America, Central America, and Haiti also suffered severe droughts. Central America and Haiti faced crisis levels of food insecurity.

2015–2016 Drought Testimonial

Name: *Prosper Chirara; male, 28 years*

Location: *Yafele Village, Ward 16, Goromonzi District, Mashonaland East Province, Zimbabwe*

Date: *December 12, 2017*

My name is Prosper Chirara. I am 28. I come from Mutare on the eastern border of the country with Mozambique. I have been married for 5 years and we have two beautiful daughters. I passed my Form Four (Ordinary Level) with 5 subjects. Unfortunately, I could not afford to proceed with my education (Advanced Level or vocational technical training) as I had wished, since my mother could not afford the school fees. My father is deceased.

  I came to Harare (the capital city of Zimbabwe) 8 years ago looking for work. I stayed for 2 years with an uncle without finding a job in Harare, upon which I decided to relocate to Goromonzi Business Centre, some 40km east of Harare, after a cousin told me I could secure part-time jobs on construction projects.

  When I came to Goromonzi 5 years ago, there was a lot of construction work. I easily got connected to some local builders who provided me with part-time labor opportunities such as digging foundations, working as an assistant builder mixing mortar, providing bricks and stones, and fetching water. I would move from one project to another quite easily and the opportunities were readily available.

  The most lucrative job I used to get was fetching water for construction projects for people building in the area. Most people constructing houses in the area have not yet sunk deep wells. So they rely on hiring people to fetch water for them from a local dam, river or community borehole. Fetching water to fill up a 200-litre drum would earn $2. I would average 5 drums a day, earning about $10 a day. With an average 5 days a week that would be $50. Multiply by 4 weeks, that would be $200. $200 just for fetching water! Then consider the other part-time jobs I would do besides that. In a good month I could make between $250 and $350 per month. That is more than what most professional civil servants earn in a month!

Back then, I used to make good earnings that saw me looking after my family very comfortably. My family was well-fed. I would also send my mother money every month to buy food and other basics. My mother is 57, and though she still able to work the fields, she suffers from high blood pressure. I also could afford the medication that she requires.

The first three years I was here (Goromonzi) things were really really good for me. I also managed to buy some household items (beds, set of chairs, bicycle, utensils, clothes etc.) for my family.

Then came the drought from 2014 to 2016! The year 2015 into 2016 was really bad. I will never forget that period. Things changed completely, and my life changed. Even now as you see me I have not recovered fully from the impacts of that drought.

That drought caused many households' deep wells to dry up. Water tables receded so much that the local dam dried off. The local river stopped flowing.

Most people who were constructing houses stopped work on their projects due to water challenges. Very few could afford to hire water trucks or supply tankers. The long and ever-present queues at the borehole ended construction projects. It would take one up to 5 hours for one's turn to come to draw water. And one was only allowed two 20-litre buckets at a time. After which one had to join the queue from the back and wait for another five or so hours. People would literally not go to sleep at night in order to take their stands in the queue. Some would wake up as early as 1 am or 2 am, upon which the queue would still be there.

My income dwindled so pathetically before my own eyes. This was worsened by the cash shortages that began in early 2016. Household incomes for casual laborers fell. For the few jobs available, the rates offered were very low.

My friend, things went tough for me. My income was just a trickle. Sometimes I would get a small job, but the "bosses" now would take a long time to pay. I don't forget the hundreds of

> kilometers I walked from one boss to another pleading to be paid. The savings that I had were cleaned up!
>
> My family struggled from that drought. Putting food on the table became a big challenge. I could not afford the most basic of groceries. What made the situation worse was that prices for food and other basic commodities were increasing regularly. My family literally "starved". I ended up borrowing heavily just to make sure I got food for my family. I am still working to clear off some of the debts I owed.
>
> All the support I used to offer my mother just evaporated. I could not afford to send my mother any money for groceries. At least my mother was among the food assistance beneficiaries in our village – so I knew she had some food from that source, though not enough! What pained me most is that I would receive calls from my neighbor informing me that my mother was not feeling well, yet I could not afford to go and see her.
>
> My friend, I am happy that the rains were good last season (2016-17). I managed to have a good crop of maize, which stocks we are still consuming and will likely last the next two months. It's not back to five years back yet, but I can make up to $50-$60 per month. At least I am happy I can manage to send my mother something, even if it's small.

Prosper's testimonial conveys so much about his struggles. He is bright and hardworking, with an academic prowess that could have got him into college if he and his family could have afforded it. Like many young Africans, the lack of opportunity led him to the capitol city, in this case Harare, where he worked hauling water. Zimbabwe is a land-locked nation just north of the Republic of South Africa. Once viewed as a shining light of Africa's future, thirty-plus years (1987–2017) under the despotic (now deposed) rule of Robert Mugabe reduced this country, which once hosted the greatest Iron Age settlement south of the Sahara Desert, to an economic shadow of its former self. Then came the El Niño–related droughts of 2014/2015 and 2015/2016.

These droughts were part of an extensive swath of multiyear dry conditions that covered a lot of our planet.

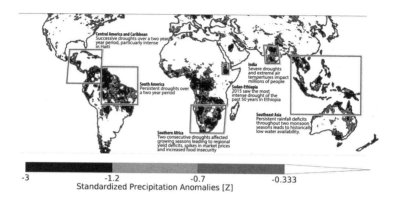

Figure 9.5 World Food Programme Analysis of multiyear El Nino impacts. Graphic by author but based on a similar figure from the WFP report El Niño: Implications and Scenarios for 2015–2016.

Analysts at the WFP analyzed the impacts of these droughts[22] and highlighted the dangerous implications of the persistent multiyear dry conditions associated with the El Niño–like climate conditions that existed from June 2014 through May 2016. To highlight these global impacts, I have produced results similar to one of their figures (Figure 9.5). This image shows two-year precipitation totals that begin in June 2014 and end in May 2016. These totals are expressed as standardized precipitation index (SPI) values. These values range from about −2 to +2. Average values have a value of zero. Exceptionally wet or dry regions will have values of less than −1.2 or higher than +1.2. In 2014–2016, areas of extreme persistent dryness stretched across many tropical regions typically associated with dry conditions during El Niños. Central America, the Caribbean, Amazonia, Sudan/Ethiopia, southern Africa, India, and Southeast Asia experienced several years of poor rain.

While conflict, civil unrest, and economic crises have played an important role in increasing food insecurity, these El Niño–associated droughts also contributed to a rise in both the

---

[22] documents.wfp.org/stellent/groups/public/documents/ena/wfp280227.pdf.

number of moderately food-insecure and extremely food-insecure people. Every year, the UN Food and Agriculture Organization (FAO) releases a report on the state of global food security. For several years (the years following the 2015/2016 El Niño), these reports highlighted a concerning increase in the number of undernourished people, with 820 million people (more than 1 in 10) facing serious caloric deficiencies in 2019.[23] Estimates of extreme food insecurity,[24] focused on identifying people facing crisis levels (or worse) of food insecurity, show an 87 percent increase since 2015, with some 88 million people likely in need of emergency assistance in 2020.

Eighty-eight million people is equivalent to the combined populations of New York, Washington, Boston, Chicago, London, Rome, Mexico, Tokyo, Delhi, Sydney, Moscow, Bogota, Hong Kong, and Shanghai. If 88 million people joined hands, they could circle the globe thirteen times. Eighty-eight million people is 1 out of every 100 people. Out of 88 million extremely food-insecure people we might have 18 million children, and about 2.7 million acutely malnourished children. That's an entire city of Chicago's worth of extremely hungry children: children with stick-thin arms.

Attributing Potential Climate Change Impacts on the 2015/2016 El Niño

If climate change made the 2015/2016 El Niño more intense, then climate change helped produce severe droughts that directly contributed to massive increases in food insecurity and human suffering. But did it? I absolutely believe the answer is "yes" – but how one answers this question may depend on your conceptual model of climate change. As discussed in Chapter 8, if one simply defines climate change as the average of a large number of climate change simulations, or the long-term trend in some variable, then the answer might be "no." For

---

[23] www.fao.org/3/ca5162en/ca5162en.pdf.
[24] fews.net/sites/default/files/Food_assistance_needs_Peak_Needs_2020_Final.pdf.

example, climate attribution analyses incorporating such a framework have suggested that climate change did NOT intensify the 2015 Ethiopia drought.[25,26]

On the other hand, however, if one begins with a conceptual model that expects climate change to increase the ocean's heat content, and that this "extra" heat content will converge in different regions at different times, then it follows that human-induced warming *did* make the 2015/2016 El Niño more intense, hurting a lot more people than it would have done in the absence of human-induced climate change. Following this basic line of reasoning, my colleagues and I have formally assessed the contribution that human-induced climate change made to the extreme 2015–2016 El Niño in three papers,[27] formally linking this anthropogenic contribution to the associated droughts and food crises in Ethiopia and Southern Africa. These food crises helped push more than 35 million people into extreme food insecurity. It is fairly well accepted by scientists that climate change *will* increase the frequency and magnitude of future strong El Niños[28]: our research emphasized that such impacts are happening *now*, hurting people like Prosper *now*.

You don't have to just believe me; you can see it with your own eyes. Figure 9.6 shows a time series of December-to-

---

[25] journals.ametsoc.org/doi/10.1175/JCLI-D-17-0274.1.
[26] www.worldweatherattribution.org/ethiopia-drought-2015/.
[27] Funk, C., L. Harrison, S. Shukla, A. Hoell, D. Korecha, et al. "Assessing the contributions of local and east Pacific warming to the 2015 droughts in Ethiopia and Southern Africa." *Bulletin of the American Meteorological Society* (December 2016): S75–S77, doi:10.1175/BAMS-16-0167.1. journals.ametsoc.org/doi/abs/10.1175/BAMS-D-16-0167.1; Funk C., F. Davenport, L. Harrison, T. Magadzire, G. Galu, et al. "Anthropogenic enhancement of moderate-to-strong El Niños likely contributed to drought and poor harvests in Southern Africa during 2016." *Bulletin of the American Meteorological Society*, 37 (2017): S1–S3, DOI. 10.1175/BAMS-D-17-0112.2.www.ametsoc.net/eee/2016/ch18.pdf; Funk C., L. Harrison, S. Shukla, C. Pomposi, G. Galu, et al. "Examining the role of unusually warm Indo-Pacific sea surface temperatures in recent African droughts." *Quarterly Journal of the Royal Meteorological Society* (2018). doi.org/10.1002/qj.3266.
[28] Cai W. et al. "Increasing frequency of extreme El Niño events due to greenhouse warming." *Nature Climate Change* (2014). www.nature.com/articles/nclimate2100.

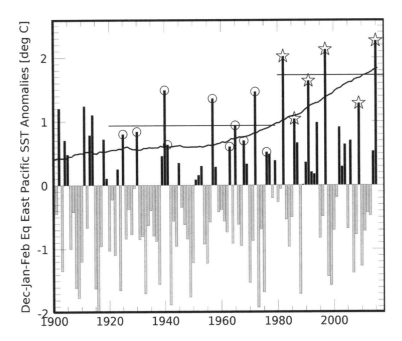

Figure 9.6 Observed December–February equatorial East Pacific sea surface temperature anomalies. Based on version 5 of the NOAA Extended Reconstruction data set averaged over the Niño3.4 region. The circles and stars mark out strong El Nino events during the 1920s–1970s and from 1981 to 2016. The horizontal bars denote the average of these events.

January equatorial East Pacific[29] sea surface temperatures. The December–February time period tends to be both the peak of El Niño events and the heart of the Southern African crop-growing season. These sea surface temperatures are commonly used to measure the intensity of El Niño. The data are shown as "anomalies" or differences from the long-term average. When these anomalies are greater than about +0.5 °C, we enter mild El Niño conditions. When temperatures are above about +1.5 °C, we are experiencing a strong El Niño.

---

[29] These equatorial East Pacific sea surface temperatures were averaged over the Niño3.4 region (170°E–120°W, 5°S–5°N).

Visually, there a few striking features of Figure 9.6. It looks as if recent ocean temperatures may be more variable, and that recent El Niño events (noted by very warm "spikes" in the time series) may be substantially warmer. El Niños typically occur once every five to seven years or so. We can isolate these El Niño events by looking at the average of the top one-in-six warm events over a given period of time. This corresponds to the typical strength of a moderate to strong El Niño event. For example, we can look at the average temperature of the six warmest anomalies between 1981 and 2016. We can also look at the average temperature of the ten anomalies between 1921 and 1980. These are the peaks in Figure 9.6.

The two horizontal bars in Figure 9.6 shows the average temperature anomaly associated with these one-in-six year El Niño events over the 1921–1980 time period and the 1981–2016 period. In this data set we find a huge difference between "new" (1981–2016) and "old" (1921–1980) El Niños. New El Niños are much warmer, about 0.8°C warmer. While other data sets give slightly different answers, they all agree on a similar, and disturbing, story. We have seen a large and statistically significant increase in the intensity of El Niño events. This magnitude of change could transform a weak-to-moderate El Niño into a strong El Niño. We will find out that that difference can be supercritical – helping substantially increase the chance of an El Niño–related drought.

But maybe the results shown in Figure 9.6 are just due to chance? We have a very limited historical record. Historically, sea surface temperature observations were taken from ships, who used the measurements to help track their location in the oceans. Before 1920, there were very few ships traveling in the eastern equatorial Pacific. Strong El Niños are also infrequent, which makes it statistically difficult to identify changes in their extremes. Furthermore, we only have one planet, so we only have a single time series to examine.

Using climate simulations, we can formally address these types of questions using climate attribution methods. Using climate models, we can create simulations over a much

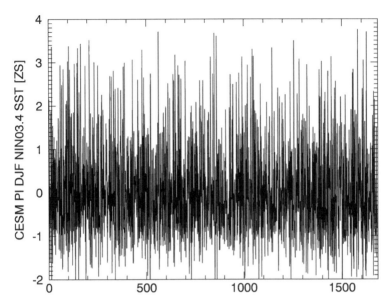

Figure 9.7 Seventeen hundred years of standardized December–February Niño3.4 sea surface temperatures from a preindustrial simulation using version 1 of the Community Earth System Model.

longer time period than a few hundred years. And, if we are interested in the behavior of a world without climate change, we can use long "preindustrial simulations." These global ocean-atmosphere model simulations have fixed low levels of greenhouse gasses and aerosols. They are often used to characterize "normal" behavior in a world without human emissions. These simulations are typically run for hundreds or thousands of years to give us a big sample to characterize 100 percent natural preindustrial climate conditions. Figure 9.7 shows standardized December–February El Niño sea surface temperature anomalies from a 1,700-year climate simulation[30] under preindustrial conditions. We do in fact see many extreme El Niños in Figure 9.7, so it turns out that it *would* be possible to see increases like that

---

[30] Broadly following the work presented in our 2017 BAMS attribution paper, www.ametsoc.net/eee/2016/ch18.pdf, I have used preindustrial simulations from version 1 of the Community Earth Systems Model.

shown in Figure 9.6 purely by chance. We do in fact find instances in a world without human-induced climate change when a thirty-year period could experience a large increase in strong El Niños. But these occurrences are very rare. With 1,700 years of modeled data to work with, we can quantify how *likely* it might be to see a period of moderate El Niños followed by a period of strong El Niños. It *does* turn out to be possible, but highly unlikely (statistically speaking).

To estimate the likelihood of the observed behavior we see in Figure 9.6 we can take the 1,700 years of data in Figure 9.7 and break out all the possible *60-year-followed-by-36-year* time periods. Then for each of these time periods we can calculate the change in one-in-six warm events between the first 60 and the second 36 years. The dashed line in Figure 9.8 shows the probabilities of changes in sequential 60- and 36-year moderate-to-strong Niño3.4 sea surface temperatures under preindustrial conditions. The distribution is symmetric and centered about zero. Sometimes we might see a shift toward cooler conditions, simply due to chance. Sometimes we might see a shift toward warmer conditions, again simply due to chance. But over time the chances of going up or down are symmetric, so, as you might expect, this distribution is centered on zero. But overall, we can use this distribution to assess what the typical range of natural El Niño fluctuations might be, and this range is surprisingly large. El Niños can be really warm (Figure 9.8, dashed line), and runs of naturally occurring warm or cold events can produce large changes in El Niño behavior just through random chance. But when we plot the observed changes in El Niño behavior (+0.8°C) as a thick vertical line on Figure 9.8, we see that the observed shift is *extremely unlikely* given the distribution of simulated changes. The gray shading in Figure 9.8 shows the region of the preindustrial distribution with changes as large as or larger than the observed +0.8°C change. This area covers only 3 percent of the distribution, indicating that a +0.8°C change would be possible but *very unlikely* under natural conditions.

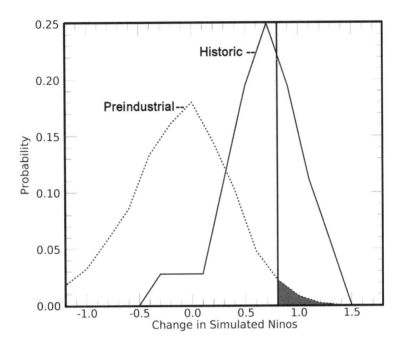

Figure 9.8 Preindustrial and historic distributions of changes in one-in-six-year warm El Niño sea surface temperature anomalies. The dashed line shows the preindustrial change distribution. The vertical black line indicates the observed NOAA Extended Reconstruction version 5 Niño3.4 temperature increase (~+0.8°C). The dark gray shading denotes likelihood given preindustrial conditions (~3%). Thick non-vertical black line indicates the probability distribution provided by the models forced with historic greenhouse gasses and aerosols.

So, how likely might a +0.8°C change be in a world *with* human-induced climate change?

To answer this question, we can use a large set of forty simulations driven with observed greenhouse gas and aerosol changes.[31] This large set of climate change simulations allows us to derive a probability distribution function showing anticipated

---

[31] Again, to broadly follow the work presented in our 2017 BAMS attribution paper, I have used forty historic simulations from version 1 of the Community Earth Systems Model.

changes between 1921–1980 and 1981–2016. Using a climate change model, we create forty worlds similar to our world, and subject these worlds to the same historic changes in greenhouse gasses, aerosols, and solar radiation as our Earth. Now, instead of just one time series like Figure 9.6, we have forty. This allows us to assess the probability of a warming similar to that seem in the observations, in a world *with* climate change. The distribution of these changes is shown with the heavy black line in Figure 9.8. The climate change simulations predict an increase in El Niño sea surface temperatures that is *very* similar to what we see in the observations (about +0.7–0.8°C). In other words, a +0.8°C change would be *very likely* in a world *with* climate change.

Putting the two pieces of this story together, we can use the preindustrial distribution (dashed line Figure 9.8) to say that it appears extremely unlikely that the observed increase in El Niño strength would have occurred in a world without climate change. We can also use climate change simulations to suggest that the observed magnitude of the increases appears to be almost exactly what we are seeing in many of our climate models and simulations (solid line in Figure 9.8). The climate change simulations indicate that the observed warming is very likely. Furthermore, the climate change models do a great job of predicting the actual rate of warming.

This allows us to say with confidence that climate change did indeed make the 2015/2016 El Niño more extreme, and exacerbated droughts like those that occurred in Ethiopia and Southern Africa. Think on that for a second. This attribution assessment has profound moral implications. Detailed assessments by humanitarian agencies identified massive increases in food insecurity and widespread economic disruption in Ethiopia and Southern Africa. Emergency assistance for 47 million people helped avert famine, but the total amount of human suffering was immense, and climate change made it worse. Which means we have all contributed to this suffering. The 2015/2016 El Niño also induced droughts and temperature extremes in many regions (Figure 9.5), and contributed to extreme tropical sea surface temperatures, fishery destruction, and coral bleaching.[32]

There is another important dimension to our attribution analysis: prediction. By finding persuasive *evidence* that climate change has *already* contributed to a large increase in the magnitude of strong El Niño events, we also deduce that climate change will likely contribute to more severe El Niño events and El Niño–related disasters in the (near) future. We are likely to continue to see more frequent very extreme El Niño events, similar to 2015/2016, accompanied by severe droughts in many tropical regions (Figure 9.5).

I am a careful scientist. I will honestly tell you we can't be *100 percent sure* that we will see a continued increase in strong El Niño behavior in the next twenty years. But the odds will not be forever in our favor. Six chambers, one bullet. Spin, click: no boom. Double or nothing with your children's future? Do we have the right to gamble the future of millions of people trying to get enough to eat? Do we have the right to gamble the livelihoods of aspiring young men and women like Prosper Chirara?

---

[32] BAMS, Explaining Extreme Events of 2015 from a Climate Perspective. journals.ametsoc.org/doi/pdf/10.1175/BAMS-ExplainingExtremeEvents2015.1; Explaining Extreme Events of 2016 from a Climate Perspective. journals.ametsoc.org/doi/pdf/10.1175/BAMS-ExplainingExtremeEvents2016.1.

10 BIGGER LA NIÑAS AND THE EAST AFRICAN CLIMATE PARADOX

Standing in front of the mid-sized Washington, DC conference room on May 19, 2016, I could feel my palms sweat. I was pretty nervous as the thirty or so important US Agency for International Development (USAID) analysts and decision-makers settled into their seats. I had never given this type of presentation to USAID before. I had flown out from Santa Barbara to talk about how climate change had made the 2015/2016 El Niño worse, and why they should all now be worried about climate change making the 2016/2017 La Niña–associated droughts more intense. There were some old USAID friends in the room – but there were also many new faces.

My job for the next thirty minutes was to try to compress about a dozen research papers' worth of insights into thirty-two slides. These insights stemmed from an intense multiyear, multipartner FEWS NET research effort that focused on understanding the decline in the East African March–May rains, and why USAID should be concerned about sequential back-to-back droughts if a La Niña emerged, as was currently predicted by many climate experts.

The stakes were high. The last time such a sequence of droughts had happened was in October–December 2010, and then in March–May 2011 a massive food crisis had broken out in East Africa. In central and southern Somalia, the violent al Shabab insurgency combined with drought, access limitations, devastated crops, and skyrocketing food prices to create a perfect recipe for famine. More than 255,000 Somalis died, and one out of ten children in southern Somalia lost their lives.

La Niña events occur when the equatorial eastern Pacific Ocean becomes very cool, and the western Pacific becomes very warm. Such events often follow strong El Niño events.

The 2011 drought was characteristic of what has been called the East African Climate Paradox.[1] This has been called a paradox because while climate models predict increases in March-to-May rains in East Africa, observational analyses originally led by me[2] and then confirmed by many others identify a disturbing tendency toward more frequent droughts. Figure 10.1 provides an updated rainfall analysis describing this tendency.[3] Since 1999, there have only been a few wet seasons: in 2010 and 2013, and the very wet 2018. Conversely, nine years had poor rainfall: 1999, 2000, 2001, 2004, 2008, 2009, 2011, 2017, and 2019. This means that the "new normal" in eastern East Africa has been a substantial or severe March-to-May drought about every other year. Such sequential rainfall deficits have had dangerous impacts, eroding resilience, economic reserves, herd size, and human health.

On May 19, 2016, I didn't know that the 2017 or 2019 rains were going to be so terribly poor, but I did know that climate experts at NOAA and the International Research Institute were predicting that a La Niña event was likely. And I did know that climate change was likely to make that incipient La Niña event more dangerous for East Africa. For many years and in dozens of research papers, FEWS NET research scientists[4] had been advancing a line of study connecting the increased frequency of East African droughts to warming in the Indian

---

[1] journals.ametsoc.org/doi/pdf/10.1175/JCLI-D-15-0140.1.
[2] pdf.usaid.gov/pdf_docs/PNADH997.pdf.   [3] blog.chc.ucsb.edu/?m=201907.
[4] Primarily Chris Funk, Park Williams, Andrew Hoell, Gideon Galu, Brant Liebmann, Shraddhanand Shukla, and Laura Harrison.

214 / Drought, Flood, Fire

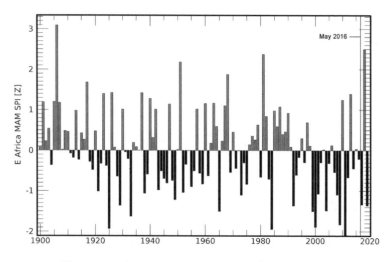

Figure 10.1 A 1900–2019 time series of standardized precipitation index (SPI) values, Based on Climate Hazards Center precipitation data, for March–April–May rainfall in central and eastern Kenya, central and eastern Ethiopia, and Somalia. SPI values have a mean of 0 and a standard deviation of 1, so an SPI value of less than −1 indicates a potentially dangerous drought. The thin vertical line depicts the date of my May 2016 presentation to USAID.

and Pacific Oceans. As first advanced in 2005,[5] our work suggested that the basic mechanism for this drying involves increases in rainfall over the warm tropical oceans to the east of East Africa. Human-induced warming in these already very warm waters can explain both the long-term decline of the East African rains,[6] and the emergent increased sensitivity to La Niña climate conditions.[7] In 2010, however, my research was still primarily focused on warming in the Indian Ocean. Then, in

---

[5] royalsocietypublishing.org/doi/full/10.1098/rstb.2005.1754.
[6] Williams, P., and C. Funk. "A westward extension of the warm pool leads to a westward extension of the Walker circulation, drying eastern Africa." *Climate Dynamics* 37.11–12 (2011): 2417–2435. https://link.springer.com/article/10.1007/s00382-010-0984-y
[7] Williams, P., and C. Funk. "A Westward Extension of the Tropical Pacific Warm Pool Leads to March through June Drying in Kenya and Ethiopia." USGS Openfile Report 1199 (2010). https://pubs.er.usgs.gov/publication/ofr20101199.

2012, important new work by Brad Lyon and David DeWitt[8] argued convincingly that the Pacific played a critical role in driving East African droughts. This research helped lead to a new set of FEWS NET papers emphasizing the interplay of warm west Pacific sea surface temperatures and cool East Pacific sea surface temperatures.[9] This work moved us beyond understanding the trend and into the realm of *prediction*. The eastern East African rains were predictable, based on the interaction of west and east Pacific sea surface temperatures. So – at my talk in May 2016 – I warned:

- FEWS NET research suggests that both El Niños and La Niñas may be becoming more intense, causing more extreme impacts in some places.
- For 2016/2017, a strong La Niña appears likely to be enhanced by a warmer west Pacific.
- The spatial pattern of La Niña impacts on East Africa intersects with some very food-insecure regions of Somalia and Ethiopia. La Niña–related drought could exacerbate existing food and water shortages.

---

[8] Lyon, Bradfield, and David G. DeWitt. "A recent and abrupt decline in the East African long rains." *Geophysical Research Letters* 39.2 (2012).
  agupubs.onlinelibrary.wiley.com/doi/full/10.1029/2011GL050337.
[9] Hoell, A., and C. Funk. "The ENSO-related West Pacific sea surface temperature gradient." *Journal of Climate* 26 (2013): 9545–9562;Hoell A., and C. Funk. "Indo-Pacific sea surface temperature influences on failed consecutive rainy seasons over Eastern Africa." *Climate Dynamics* (2013): 1–16. DOI: 10.1007/s00382-013-1991-6; Liebmann, B., M. Hoerling, C. Funk, R. M. Dole, A. Allured, et al. "Understanding Eastern Africa rainfall variability and change." *Journal of Climate* (2014); Shukla, S., A. McNally, G. Husak, and C. Funk. "A seasonal agricultural drought forecast system for food-insecure regions of East Africa." *Hydrology and Earth System Sciences Discussions* 11.3 (2014): 3049–3081; Shukla S., C. Funk, and A. Hoell. "Using constructed analogs to improve the skill of March-April-May precipitation forecasts in equatorial East Africa." *Environmental Research Letters* 9.9 (2014): 094009; Funk, C., A. Hoell, S. Shukla, I. Bladé, B. Liebmann, J. B. Roberts, and G. Husak. "Predicting East African spring droughts using Pacific and Indian Ocean sea surface temperature indices." *Hydrology and Earth System Sciences Discussions* 11.3 (2014): 3111–3136; Funk, C., and Hoell A. "The leading mode of observed and CMIP5 ENSO-residual sea surface temperatures and associated changes in Indo-Pacific climate." *Journal of Climate* 28 (2015): 4309–4329.

Figure 10.2 Pictures from Somaliland, taken by James Firebrace and forwarded by James Magrath on November 15, 2016.

By October 2016, the La Niña had arrived. A cold eastern Indian Ocean, a warm Indo-Pacific, and a cool eastern Pacific created perfect conditions for a terrible East African drought. Following the end of the 2016 El Niño, west Pacific temperatures jumped to extremely high levels. On October 19, the Climate Hazards Center issued a pessimistic forecast for the October-to-December 2016 season via our group's just-created blog.[10] On November 8 and November 16, we added blogs updating this forecast and documenting the high likelihood of poor pasture and crop conditions.[11]

Then on November 15, 2016, my friend John Magrath, who worked with Oxfam, forwarded me the following from James Firebrace, who had been bravely engaged in on-the-ground field assessments of conditions in northern Somalia. Figure 10.2 shows images from James Firebrace's original e-mail. John's message went as follows:

> **From John Magrath**
> *Fw: Drought in Somaliland - Hi Chris: we've just received some disturbing photos + information on the humanitarian situation - thought you ought to know…*
>
> *You getting any similar info? If not, can you raise concerns with people you know? You'll see these pics are not sourced but they were sent to James Firebrace, a reliable contact. The info about population movements is pretty disturbing.*
>
> *Best,*
John
>
> **From James Firebrace**
> *Richard, Beccy, Emma and Debbie*
> *This is the latest I have picked up from Somaliland: The Deyr rains completely failed in the eastern regions of Somaliland. Many of those residing in the areas most hit by the drought have migrated elsewhere. Those that were able have migrated to an area called Oodale (HAWD), where there were rains recently, its near Buuhodle. While some have migrated all the way to Garawe (Puntland). These two groups are easy to reach as they have congregated in these two areas, and include roughly 300,000 people. These people are at a high risk, due to shortages of food, water and shelter from a harsher than usual winter. Even worse are those that did not have the means to move out of the drought stricken regions, because they are much*

[10] blog.chc.ucsb.edu/?m=201610. [11] blog.chc.ucsb.edu/?m=201611.

> harder to reach. Please do your best to raise awareness about the current situation in eastern Somaliland.
> Also attached are the images sent to me ...
>
> Best wishes
> James

As you can imagine, this really got my attention. So in early December I worked very hard with my colleagues to put together a statistical forecast for the March-to-June East African long rains, which we released on December 5.[12]

On December 9, James Firebrace wrote to me from Somalia:

> Attn Chris Funk, Climate Hazards Group
> Dear Chris
> (cc relevant Oxfam staff)
> You may recognize my name – John Magrath of Oxfam sent on the earlier photos I had of the effect of the current drought in Somaliland.
> I have now been in Somaliland for close to 3 weeks (on behalf of the President and National Drought Committee) and have made a major tour of the most drought affected areas to the east. There have been massive loss of livestock in the areas I visited (over 70% of herd size across all the interviews I made), and increasing dependence on (as yet inadequate) water trucking and food aid. We are now entering traditional dry season (Jilaal, Dec to March) and looking ahead and one can foresee much greater loss of livestock, as well as human hardship, to the point where restocking herds is not going to be easy with serious implications for Somaliland's economy.
> Older people are insisting that this is the worst they've seen – worse than what they call the Long Tail (Daba Dheer) drought of 1974-5 which led to a massive resettlement programme to the South, and worse than the 'Red

[12] blog.chc.ucsb.edu/?m=201612.

> Dust' (Sig'aas) drought of the mid 1950s. What makes it different this time is the sheer extent of the drought and lack of significant areas of accessible pasture to turn to (even across the border in the Somali Region of Ethiopia). Those limited areas that did obtain a little rain (and hence pasture) were rapidly overgrazed by pastoralists seeking to save their herds, with the result that their animals ended up in even worse condition.
>
> Forgive me for asking you a few questions. If you have time to respond, even a short comment would help make judgements for the best course of action looking forward.
>
> 1. How do you see the outlook for the next Gu rains (April to June 2017) for Somaliland (and the Somali Region of Ethiopia to the south much of which is largely the same region in nomadic terms)? Can one at least get a feel for whether the recent run of poor rains might now break?
> 2. Are you able to comment on the likely link between the severity of this current drought and human-induced climate change? I appreciate this is a complex topic. How far can one justifiably go in saying that climate change is making the situation worse?
>
> > ...
> > I am back in UK on Monday and would happily set up a Skype chat if you would like further briefing on my findings. In any event I can copy you into my final report if that is useful.
> >
> > Kind regards
> > James Firebrace
> > Director, JFA

In answer to this first question, I was able to write back and send James a link to our extensive December 5 blog post. This post and several more that followed all called for a below-normal outlook for the 2017 long rainy season. On January 16, 2017, FEWS NET and Somalia Food Security and Nutrition Analysis Unit issued a joint alert calling out the high combined risk generated by the poor 2016 and anticipated poor 2017 rainy seasons.[13] During this period, we were also networking and sharing information with our partners in Europe at the Joint Research Center, and in February the Food and Agriculture Organization of the United Nations and the United Nations World Food Programme issued a joint alert citing our outlook for below normal rains in 2017[14]:

> *Humanitarian partners should urgently prepare themselves to scale up their interventions in response to food insecurity levels and food insecure population numbers in Somalia and neighbouring regions, which are likely to be at their highest levels since the 2010-2011 disaster.*
>
> *Based on the drought impact evidence included in this statement and on the more detailed food security and nutrition information included in the multi-agency assessment released on 2 February 2017, the following actions are of highest priority:*
>
> - *provision of urgent and substantial food assistance to help currently food-insecure households and support the most affected livelihoods. Response activities to be planned on the detailed IPC analysis outcomes that were issued on 2 February 2017;*
> - *updating of emergency response for agro-pastoral communities, intensification of advocacy and resource mobilisation to address the impact of an extended post-2016 "Deyr" harvest lean season. For pastoral communities, provide updates of contingency plans to face continued severe shortages in livestock forage and drinking water;*
> - *continued close monitoring of the dry January to March "Jilaal" season and the next "Gu" rainfall season to inform decision-making on programming and targeting*
> - *increase awareness of the need for a regional approach to address the effects of drought that are*

---

[13] www.fsnau.org/downloads/FEWS-NET-FSNAU-Somalia-Alert-2017-1-16.pdf.
[14] fews.net/east-africa/somalia/special-report/february-21-2017.

becoming more frequent and intense, and to ensure adequate access to populations in need of assistance.

In 2016 and early 2017, our understanding of the role played by climate change – how extremely warm west Pacific sea surface temperatures have triggered numerous frequent and predictable droughts – helped motivate a timely and effective response in late 2016 and early 2017. Additional humanitarian assistance for about half a million people was already reaching Somalia in December 2016, with the amount doubling again in January, and again in February–March. The March-to-May rain season was very poor, and the humanitarian crisis extensive, disruptive, and painful. But we did not see the terrible outbreak of famine that occurred in 2011: a disaster on that scale was averted.

We can see the influence of the improved 2016–2017 interventions in time series of sorghum prices in Baidoa, a major food-insecure city in Somalia (Figure 10.3). Between July 2010 and June 2011, food prices in Baidoa climbed by more than 300 percent, rising from about 5,000 shillings per kilogram to 18,000 shillings per kilogram. In early 2017, humanitarian assistance was helping stave off such a meteoric increase, so that between June 2016 and July 2017, prices only increased from

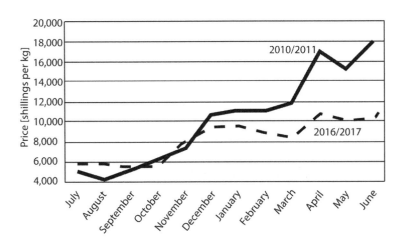

Figure 10.3 Sorghum prices in Baidoa, Somalia.

around 6,000 to 10,000 shillings per kilogram. While our forecasts were just one piece of a complex early warning tableau, understanding the dangerous relationship between human-induced warming in the West Pacific and drought in East Africa helped us provide valuable predictions.

Height and Wind Fields

This section will be more interesting if you know a little bit about atmospheric height and wind fields. The broad strokes of their relationships are pretty easy to understand. Scientists like to analyze atmospheric motions on "pressure fields." These are the paths of motion an airplane or hot weather balloon would follow. At the surface of the Earth a barometer measures about 1,000 hecto-Pascals (hPa) – this is the weight of air above you. The atmospheric maps shown in Figure 10.4 are taken at 200 hPa – when 20 percent of the atmosphere is above you and 80 percent below.

As you glide along in a weather balloon, your actual height, in meters, will increase or decrease depending on the temperature of the air below you. This measurement gives us maps of atmospheric highs and lows that we see on weather maps.

In the tropics, near the equator, winds blow from highs to lows.

In the extra-tropics, however, the spinning of the Earth works a strange magic, turning winds to the left in both the Northern and Southern hemispheres, so that they flow parallel to the height fields.

A west–east equatorial sea surface temperature gradient induces the first effect. A north–south extra-tropical gradient induces the second. These work together to create extreme El Niños and La Niñas.

Did Climate Change Make the 2017 East Africa Drought More Intense?

A key aspect of our successful forecasts of the 2017 East African drought was an increased (and increasing) understanding of how climate change is interacting with natural La

Niña variations in a "positive" but dangerous way. Positive in this sense means "amplifying," i.e., interacting in ways that increase the dynamic consequences of cool tropical East Pacific sea surface temperatures. While the link between El Niño and climate change is pretty straightforward – climate change makes tropical East Pacific sea surface temperatures warmer – the La Niña interaction is subtler. When cool east Pacific sea surface conditions are surrounded by very warm west Pacific waters, the influence of the cool east Pacific waters is increased. And this can explain why recent La Niñas are more likely to produce East African droughts.[15]

As shown in our 2018 paper in the *Quarterly Journal of the Royal Meteorological Society*, warming of the western Pacific has produced changes that increase the atmosphere's response to La Niña–like cool east Pacific sea surface temperatures. Figure 10.4 shows the *change* in top-of-the-troposphere height fields between "new" 1981–2017 La Niñas and "old" 1921–1980 La Niñas. The dark shading areas correspond with higher atmospheric heights. These correspond to the "high-pressure" patterns you might see on a weather map on your TV evening news. The top of the troposphere is about 12 kilometers up, at about the level where you see a thunderstorm flatten out into an inverted anvil. As the atmosphere warms, it expands, producing more high-pressure patterns. So almost everywhere in Figure 10.4 we see dark shading indicating these increases. Note the important exception – the subtropical eastern Pacific. Here, where cool La Niña sea surface temperatures do their deeds, we find a characteristic upper-level response. Twin upper-level cyclonic circulation anomalies (marked with black ovals and L's for "lows") describe the classic La Niña upper-level response to cool tropical ocean conditions. Wind anomalies blow counterclockwise around these upper-level lows.

---

[15] Funk, C., L. Harrison, S. Shukla, C. Pomposi, G. Galu, et al. "Examining the role of unusually warm Indo-Pacific sea surface temperatures in recent African droughts." *Quarterly Journal of the Royal Meteorological Society* (2018). doi.org/10.1002/qj.3266.

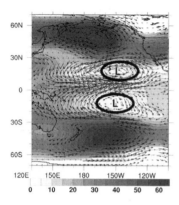

Figure 10.4 Upper-level height changes in new versus old La Niña events. From Funk et al. 2018. The ovals marked with L's indicate the upper-level cyclonic low-pressure anomalies characteristic of La Niña-like climate conditions. The dark regions over the northern and southern Pacific Ocean represent areas where warming in the western Pacific has resulted in increasing upper-level high-pressure anomalies. The interaction of these low and high-pressure anomalies drives rapid upper-level winds (arrows) that converge along the equator between 170°E and 150°W. This enhances the Walker circulation and La La Niña-like climate variations.

Now comes the interesting part. Around these lows, and especially across the northern and southern Pacific, we find very big (50 meter) increases in upper-level height fields. The intensification of these height fields is associated with long-term warming influences in the western and extra-tropical central Pacific. The warming ocean warms the air, which becomes less dense and stretches, increasing upper-level height fields. This region of high geopotential heights (shown with dark shading in Figure 10.4) wraps hand-in-glove around the La Niña upper cyclones, increasing atmospheric disruptions associated with La Niña events. Winds (shown with arrows in Figure 10.4) respond to the gradient (difference) between the low- and high-pressure cells. Just to the north of the northern La Niña low and just to the south of the southern La Niña low we see super-strong height gradients and very rapid westward wind anomalies at about 30°N and 30°S.

Such anomalies do not bring glad tidings to California and the southern United States because they blow right into the teeth of North America's storm tracks. Traveling farther east, these anomalies get turned toward the equator by rising height fields to the east of China and Australia. This is caused by the warming of the west Pacific. Converging and turned to the east along the equator, we find a very strong enhancement of the Walker Circulation. This enhancement *is not just due* to the local influence of cool or neutral east Pacific Ocean temperatures, but rather relies on the interaction of the tropical East Pacific lows with the surrounding high pressures, and the overall counterclockwise turning circulation patterns.

One way to think about Figure 10.4 is that the dark shaded areas are doing what we expect a warming atmosphere to do – expand, which makes the upper-level height field get higher. When there is a La Niña, there is a natural "warming hole" associated with La Niña's characteristically cool east Pacific waters. And over these naturally cool waters we find natural upper-level lows (marked with ovals and L's in Figure 10.4). This interaction can explain a greater sensitivity to naturally occurring La Niña conditions.

This probably seems like mega-geek-a-nomics to you. But these insights helped save people's lives in 2017. There are many climate scientists who think that the East African March–May rains are very hard to predict. They don't appreciate that climate change has upped the ante, making the region much more susceptible, in predictable ways, to west Pacific warming and East Pacific cooling. Our appreciation of this sensitivity allowed us to start making successful predictions of the 2017 March-to-June East African drought in early December 2016,[16] based on November 2016 data. These forecasts helped motivate effective and early intervention. When the March and April rains in Southern Somalia failed again, assistance was already arriving.

---

[16] blog.chc.ucsb.edu/?m=201612.

Did Climate Change Make the 2017 March-to-May Drought More Intense?

Did human-induced warming in the west Pacific increase the probability or severity of the 2017 East African drought?

Insights into recent East African droughts can be gained by looking at simple "composites" of sea surface temperature like those shown in Figure 10.5. Drought composites are produced by averaging conditions during drought years. Figure 10.5 shows the typical state of the world's oceans during recent East African droughts. Areas with dark grey shading are exceptionally warm. The black boxes over the western Pacific define a region I refer to as the Western V region. Warm waters in this region can produce a La Niña–like response associated with dangerous dry conditions over Somalia, Kenya, and southeastern Ethiopia. March-to-June East African droughts are associated with very warm Western V sea surface temperatures. The Western V region is delineated with black boxes in Figure 10.5. When the Western V region becomes very warm, you see circulation anomalies like those plotted in Figure 10.4.

The time series shown in Figure 10.6 display the magnitude of Western V sea surface temperatures in 2017. The top

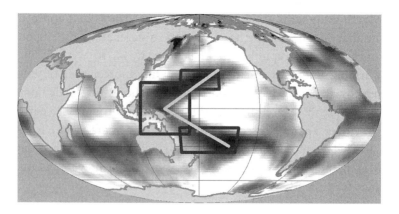

Figure 10.5 Standardized sea surface temperature anomalies associated with recent March–June East African droughts.

227 / Bigger La Niñas and the East African Climate Paradox

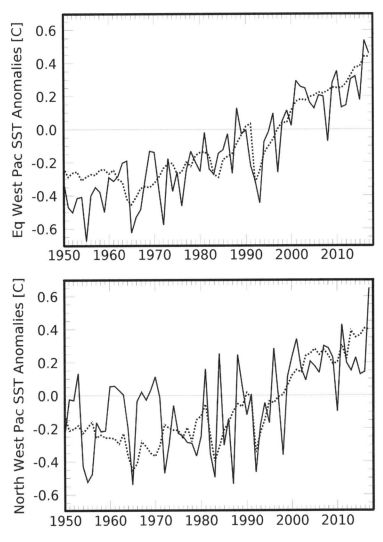

Figure 10.6 Time series of observed and modeled March-to-June sea surface temperature anomalies from regions of the Western V area. The top panel shows values for the equatorial West Pacific. The bottom panel shows values for the Northwest Pacific. Solid lines correspond to NOAA Extended Reconstruction version 5 observations. The dashed lines show the average warming predicted by the large Phase 5 Coupled Model Intercomparison Project (CMIP5) climate change ensemble.

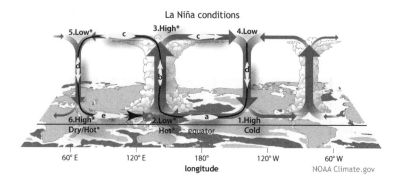

Figure 10.7 La Niña modifications to the Walker Circulation along with climate change enhancements.

panel shows conditions in the West Pacific. The bottom panel shows conditions in the Northwest Pacific. Both regions were incredibly warm, with either the warmest or second-warmest sea surface temperatures on record. Estimates of the influence of climate change, shown with dashed lines in Figure 10.6, indicate that human-induced warming contributed about 0.6°C of warming.

Warm ocean conditions in the equatorial western and northwestern Pacific act together to create more extreme La Niña–like climate conditions and droughts over East Africa and Southwestern North America. To describe this dangerous partnership, I have once again adapted a Walker Circulation schematic developed by NOAA[17] to explain how La Niña enhances the global east–west Walker circulation, producing drier conditions over East Africa (Figure 10.7). While describing these responses will require a few paragraphs, remember that the lives and livelihoods of people like Eregae Lokeno Nakali and Aita Eregae Nakali (Figures 1.1 and 1.2) may depend on these insights.

The numerals (1–6) in Figure 10.7 indicate characteristic environmental conditions during La Niña events. Several of these annotations have been marked with asterisks to indicate

---

[17] https://www.climate.gov/sites/default/files/Walker_LaNina_2colorSSTA_large .jpg.

climate change enhancement (discussed below). The associated atmospheric responses are noted with letters (a–f).

La Niña events begin with cooler-than-normal East Pacific sea surface temperatures (1) and warmer-than-normal West Pacific sea surface temperatures (2). Cool conditions over the East Pacific produce higher atmospheric pressures because the air there is denser. Over the West Pacific the air is less dense. During La Niña events the low-level winds (a) flow from the denser East Pacific toward the less-dense region around Indonesia.

These warm moist winds (a) and the warm sea surface temperature conditions (2) warm the air above Indonesia. The warmer atmosphere becomes less dense and stretches upward (b), producing higher than normal pressures in the upper atmosphere (3). This anomalous upper-level high (3) rises above the East Pacific (4) and East Africa (5). Like a skier on the top of a hill, air parcels at this location (3) have extra potential energy. The release of this energy sets up east–west upper-level wind anomalies blowing away from Indonesia toward both East Africa and the East Pacific (c). Over East Africa, in the upper atmosphere (5), these wind anomalies converge, slow, and sink (d), producing hot, dry high-pressure conditions at the surface (6). This surface high-pressure cell can also slow the westward flow of moisture into East Africa (e). So La Niña conditions can produce dry conditions in East Africa by both producing stable sinking atmospheric conditions (d) and disrupting the onshore flow of moist air from the Indian Ocean (e).

Climate change enters into our story by enhancing West Pacific sea surface temperatures (Figure 10.6) and the associated West Pacific upper-level high pressure cells. These correspond to 2 and 3 in Figure 10.7. Climate change has produced strong upward trends in these fields. When naturally occurring cool East Pacific sea surface temperatures (1) or low East Pacific upper-level pressures arise during "new" La Niñas, these normal conditions lie alongside anthropogenically enhanced conditions in the West Pacific. This combination increases all of the La Niña atmospheric responses (a, b, c, d, e), leading to hotter and drier conditions over East Africa.

Did climate change contribute substantially to the 2017 Western V sea surface temperatures and associated extreme East African drought?[18] To answer this question we can contrast two large collections of climate change simulations. One collection has been forced with observed greenhouse gas and aerosol concentrations. This represents a world with climate change. Another collection has been forced with conditions representative of preindustrial conditions. This represents a world without climate change. In both cases, the simulations were developed using version 1 of the Community Earth Systems Model.

This collection of simulation results can be expressed as probability distribution functions (PDFs). Figure 10.7 shows these results. One PDF, shown with a dashed line, represents the distribution of sea surface temperatures in a world without climate change. Another PDF displays the probability distributions in a world with climate change.

The difference in equatorial West Pacific temperatures between the world with and the world without is fairly modest: +0.5°C (Figure 10.8, top) but statistically significant and dynamically important. This region is so warm that a half-a-degree Celcius shift can enhance the West Pacific precipitation substantially, increasing rising atmospheric motions near Indonesia (b in Figure 10.7). Interestingly, the shift for the Northwestern Pacific is substantially larger: +1.1°C (Figure 10.8, bottom). There is a tendency for some scientists to discount warming in this region as natural decadal variability, but both the observations and the climate change simulations indicate *substantial* human-induced warming (bottom panel Figure 10.6, bottom panel Figure 10.8).

Would you like to know how likely the observed West Pacific sea surface temperatures would have been in a world

---

[18] This analysis is based on the 2018 study for the annual BAMS Explaining Extreme Events special issue, Funk C. et al. (2018) Examining the potential contributions of extreme 'Western V' sea surface temperatures to the 2017 March-June East Africa Drought (2018), *Bull. of Am. Met. Soc.*, S1–S6, DOI:10.1175/BAMS-D-18-0108.

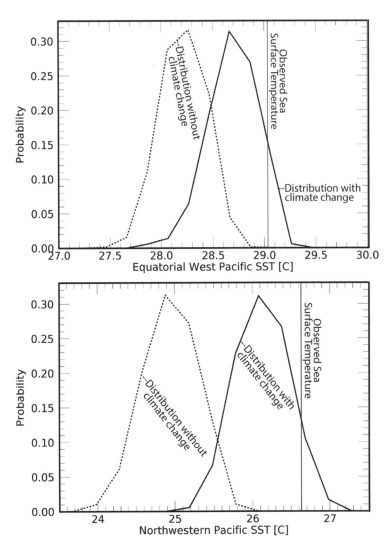

Figure 10.8 Western V SST attribution analyses. The left panel shows probability distributions for Equatorial West Pacific sea surface temperatures for a preindustrial world without climate change and current world with climate change. The observed sea surface temperature value is shown with a vertical black line. The right panel shows a similar plot for the Northwestern Pacific.

without climate change? One way to answer this question is to calculate the Fraction of Attributable Risk (FAR).[19] FAR is defined as 1 − [Probability in a world without climate change]/[Probability in a world with climate change]. We can calculate FAR values using the observed 2017 conditions in the western and northwestern Pacific (vertical black lines in Figure 10.8). Thus the FAR values were 1. In plain language, this means that such an event would have been *impossible in a world without human-induced climate change*.

But did the exceptionally warm Western V sea surface temperatures contribute to a 2017 drought over East Africa? Our 2018 attribution study explored this question by contrasting rainfall simulations from a world with and world without climate change. Using PDFs like those shown in Figure 10.8, we performed a formal attribution study by assessing the probability of drought in East Africa. In the world with climate change, Western V warming produced a La Niña–like response that dried East Africa. This means that human-induced warming in the Western V region enhanced the probability of drought over East Africa. This formal attribution study contrasted a world with and world without climate change. These PDFs indicate that climate change *doubled* both the *magnitude* and *probability* of the drought. So: *warm Western V conditions doubled the probability of the 2017 East African drought*.

Putting the Pieces Together

These results may seem complicated, and might even seem somewhat pedantic to you, but please let me explain why they are anything but. Food crises in Kenya, Ethiopia, and Somalia are strongly related to the rainfall deficits, made more intense by climate change. FEWS NET estimates indicated that between 5 and 10 million Somalians, 2.5–5 million Ethiopians, and 0.5–1 million Kenyans faced acute food insecurity in early

[19] Allen Myles, "Liability for Climate Change," *Nature* 421.6926 (2003): 891–892.

2018.[20] The results presented here demonstrate that climate change contributed substantially to this severe food insecurity. Good science can go beyond the explanation of past events: it can support *prediction* and *remediation*. The science presented here informed accurate climate forecasts that helped inform and motivate effective early humanitarian responses. The 2016 October-to-December East African drought was predicted in mid-October.[21] The following March–May 2017 East African drought was predicted on December 5, 2016.[22] This climate information, along with field reports and satellite observations, was used collectively by many international partners who worked together to provide effective early warning (Figure 10.3).

By warming the tropical Indo-Pacific, climate change is having dangerous impacts, increasing the frequency of East Africa's droughts. But by understanding how these exceptionally warm sea surface temperatures induce droughts[23] we can improve our ability to anticipate these impacts, and do something to alleviate at least some of the human suffering. By combining appropriate conceptual models of climate change with the tremendous power of the latest generation of climate models, we can provide predictions of La Niña-related droughts at very long leads. As another La Niña formed in 2020, we pushed our forecast window back to June, warning of likely back-to-back October-to-December and March-May eastern East African rainfall deficits.[24]

---

[20] fews.net/sites/default/files/documents/reports/July%202017_FAOB_final.pdf.
[21] blog.chc.ucsb.edu/?m=201610.   [22] blog.chc.ucsb.edu/?m=201612.
[23] rmets.onlinelibrary.wiley.com/doi/full/10.1002/qj.3266
[24] https://blog.chc.ucsb.edu/?p = 757 https://blog.chc.ucsb.edu/?p = 937 Funk, Chris. "Ethiopia, Somalia and Kenya face devastating drought." Nature 586.7831 (2020): 645–645.

11 FIRE AND DROUGHT IN THE WESTERN UNITED STATES

November 2018, Saturday afternoon, deep in the brushy chaparral in the woods behind my house. Temperatures sizzle. Sweat beads and drops from by body. Bone-dry, steel-strong, the thin manzanita branches resist and rebound, sending the blade leaping back with deadly intent. Rebuked, violently returned and twisted, the flat of the machete bounces off my forehead.

Some things will really make you stop and think, and a machete bouncing off your forehead is one of those things. It was pretty amazing. There were very few areas on that machete that were not razor sharp or lined with serrated saw teeth. I was totally isolated and alone, clearing a trail in the woods behind my house at the top of San Marcos Pass, which sits above Santa Barbara. I had been incredibly lucky.

The week prior had been surreal. Undergraduates on the University of California, Santa Barbara campus strolled about barely clad in 90°F mid-winter weather. The vegetation around my house was *very* dry. Though the weather was hot, rapid winter winds whipped our California coast. Drought combined with blazing temperatures had sucked all the water from the ground and vegetation. When air temperatures increase under dry conditions, relative humidity values decline,

235 / Fire and Drought in the Western United States

Figure 11.1 El Machete peligroso.

leading to drier vegetation. Plants' photosynthesis transforms carbon dioxide and sunlight into life-sustaining sugars. Drawing in the CO_2, however, means that plants must open tiny little doors in their leaves. These doors, called stomata, also let out water vapor as they draw in carbon dioxide. So, plants in arid and semi-arid regions face an existential question. How can they eat (photosynthesize) without dying of thirst, becoming weak and susceptible to the increases in disease and pest invasion that often accompany severe water stress? For California, and much of the southwestern United States, the answer to this question is "not very well." Since 2010, the U.S. Forest Service estimates that some 129 million trees have died in California's national forests "due to conditions caused by climate change, unprecedented drought, bark beetle infestation, and high tree densities."[1]

Concerned by the dry warm conditions, on November 6, 2018, I participated in a radio interview on a local radio show, "Community Alert – Not If, but When – Positive Preparation for Disaster"[2] – hosted by my friends and neighbors Mike Williams and Ted Adams. Mike and Ted are very active within the volunteer firefighter community. I periodically talk on their show, providing a climate perspective to current events.

The Southern California Geographic Coordination Center tracks climate and fire risk conditions, and in their early November Outlook[3] they described California's current situation. California had been exceptionally dry over the past twelve months, especially Southern California, with 30 percent less rain over much of the state, and 50 percent in the southernmost areas.

Fire risk, however, is also a function of temperatures, which act to increase the general level of aridity by increasing the atmosphere's water-holding capacity. Warmer air temperatures create more atmospheric "pull" for water vapor, drawing proportionally more moisture from plant stomata and bare soil. For example, my friend Park Williams and his coauthor John Abatzoglou in their paper, "Impact of anthropogenic climate change on wildfire across western US forests,"[4] show that "human-caused climate change caused over half of the documented increases in fuel aridity since the 1970s and doubled the cumulative forest fire area since 1984." Figure 11.2 is taken from their paper. There is a very strong and clear relationship between increasing aridity and forest fire area in the Western United States.

Being the data geek that I am, I wanted to include an updated time series of US wildfire extent in this chapter.

[1] https://www.fs.fed.us/psw/topics/tree_mortality/california/documents/DroughtFactSheet_R5_2017.pdf.
[2] https://www.wildlandresidents.org/community-alert/.
[3] https://gacc.nifc.gov/oscc/predictive/weather/index.htm.
[4] https://www.ncbi.nlm.nih.gov/pmc/articles/PMC5081637/.

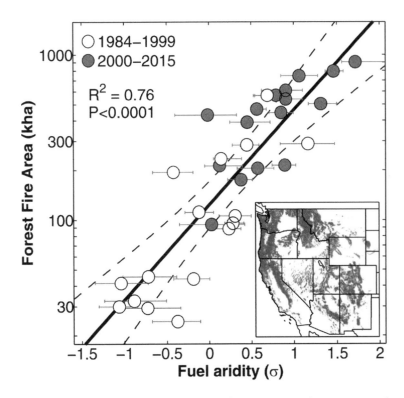

Figure 11.2 Abatzoglou and Williams's Figure 1 from "Impact of anthropogenic climate change on wildfire across western US forests." The x-axis shows a measure of aridity, taking into account both precipitation and temperature. The vertical y-axis shows the total area burned in each year since 1984.

National wildfire extent totals for 1980–2019, obtained from National Interagency Fire Center, are shown in Figure 11.3. The extent of US wildfires is increasing rapidly. I have included on this plot a simple regression-based trend estimate for 2030. If the current trend in wildfire extent continues, typical wildfire extent values in 2030 will be around 9.9 million acres – similar to the worst years to date (10 million in 2017 and 10.1 million in 2015). This is a very crude approximation, but the general point is valid – we should expect wildfires to get bigger as the western US warms and atmospheric water demand increases.

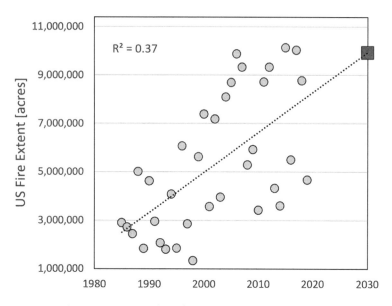

Figure 11.3 Annual US fire extent, in acres. Trend shown with a dashed line ($R^2 = 0.37$), along with a regression estimate for 2030 wildfire extent (dark square at upper-right).

Figure 11.4 shows some data that I discussed with Mike and Ted on November 6, 2018. I was very worried about the potential for yet another very bad California fire winter. This was based on both my direct experience of the dry vegetation conditions and my examination of the current climate conditions. Figure 11.4 shows ten-year averages of annual October–September rainfall, air temperatures, and standardized Palmer Drought Severity Index (PDSI) values. The top panel shows rainfall, which exhibits strong inter-decadal variations and perhaps some of the driest conditions on record. The middle panel, however, shows an unequivocal shift toward much warmer conditions. Since the start of the twentieth century, we see an annual increase in annual temperatures of about **4 degrees** Fahrenheit. Note that between approximately 2005 and 2018, California experienced a very rapid 1-degree increase. These rising temperatures are associated with increased moisture losses due to increases in evaporation.

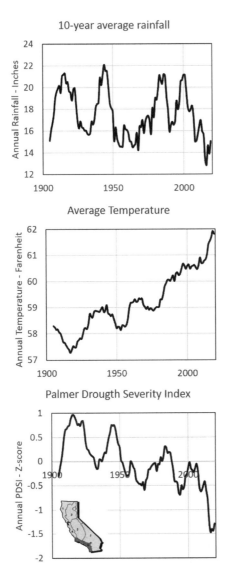

Figure 11.4 Ten-year averaged annual rainfall, air temperatures, and standardized Palmer Drought Severity Index values for the South Coast Drainage (shown in the map on the bottom left).

One measure of the balance between precipitation supply and loss through evaporation is the Palmer Drought Severity Index (PDSI). Negative PDSI values indicate an increase in evaporation losses relative to precipitation. According to the PDSI (bottom panel Figure 11.4), we have seen a major shift toward drier conditions in my neighborhood, the South Coast Drainage area. This large decrease in relative moisture supplies creates conditions conducive to fires. As these conditions persist into the early part of California's rainy season, the risk of fire can actually increase as more windy weather creates a potential for red-flag conditions (red-flag conditions indicate extreme fire risk).

For California, conditions conducive to recent droughts and fire are often accompanied by a Ridiculously Resilient Ridge – a moniker coined by Daniel Swain[5] to describe a persistent high-pressure cell over the northeast Pacific region. This high-pressure ridge can block storms and produce dry hot subsiding air over California. As pointed out by the work of Swain and others, this ridiculously resilient ridge can help explain California's recent demise into drier-hotter conditions. As context, Figure 11.5 shows a long time series of annual November–October upper-level heights near the Ridiculously Resilient Ridge. These data extend from 1948/49 through 2018/2019. Since 2012, six out of seven years (2013, 2014, 2015, 2016, 2018, 2019) have exhibited above-normal "ridging" that have helped produce dry warm weather in California. Such were conditions in early November 2018, when I experimented with deep-forest auto-lobotomy techniques (Figure 11.1) and then sat down to talk with Mike and Ted. On air, we shared our collective concern – that winds would whip up before the winter rains came. These winds combined with extremely dry fuel conditions, set the perfect stage for fire.

Unfortunately, these fears were verified. Over the next few days much of California experienced wind gusts of up to 55 miles per hour accompanied by extremely low relative humidity (humidity less than 10 percent). At sunrise on the morning of November 8, the Camp Fire, California's deadliest

---

[5] Dan has a fantastic California weather blog: https://weatherwest.com/.

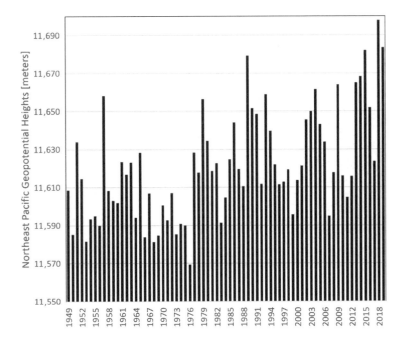

Figure 11.5 Annual November–October upper-level height anomalies over the northeast Pacific.

and costliest conflagration, broke out in Butte County[6] (Figure 11.6). At eight o'clock in the morning the fire entered the town of Paradise. Wind speeds that day reached 50 miles per hour, and the fire spread very rapidly. By November 10, more than 6,700 structures were destroyed; by the 20th, almost 20,000 structures had been destroyed.[7] Ultimately, eighty-six people died. The fire was the costliest disaster of 2018[8] ($16.5 billion), according to the international insurance agency Munich Re, and super-deadly. Two more fires broke out on November 8 – the Woolsey and Hill Fires – in Los Angeles and Ventura counties, just south of Santa Barbara, ultimately destroying an additional 1,647 structures.

---

[6] https://en.wikipedia.org/wiki/Camp_Fire_(2018).
[7] http://cdfdata.fire.ca.gov/ .
[8] https://www.usatoday.com/story/news/2019/01/08/natural-disasters-camp-fire-worlds-costliest-catastrophe-2018/2504865002/.

Figure 11.6 Landsat 8 image of the Camp Fire in northern California, taken on November 8, 2018. By NASA (Joshua Stevens) – NASA Landsat 8 Operational Land Imager.

In California, the 2017 and 2018 wildfire seasons were massive, deadly, and costly. In both years, the fire season stretched from April to December, and across much of the state. The overall size of these wildfires has risen dramatically from the late 1990s[9] (Figure 11.7). The US National Oceanic and Atmospheric Administration tracks disasters with costs over a billion dollars.[10] Here is their synopsis of the 2017 and 2018 California fire seasons.

2017

A historic firestorm damages or destroys over 15,000 homes, businesses, and other structures across California in October. The combined destruction of the Tubbs, Atlas, Nuns, and Redwood Valley wildfires represents the costliest wildfire event on record, also causing forty-four deaths. Extreme wildfire

---

[9] Data from https://www.fire.ca.gov/stats-events/.
[10] https://www.ncdc.noaa.gov/billions/events/US/1980-2018.

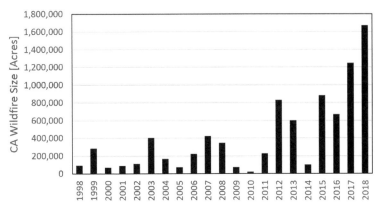

Figure 11.7 Total size of California wildfires. Values for 2019 were not available in early 2020.

conditions in early December also burned hundreds of homes in Los Angeles. Numerous other wildfires across many western and northwestern states burn over 9.8 million acres exceeding the ten-year annual average of 6.5 million acres. Montana in particular was affected by wildfires that burned in excess of 1 million acres. These wildfire conditions were enhanced by the preceding drought conditions in several states.

2018

In 2018, California has experienced its costliest, deadliest, and largest wildfires to date, with records back to 1933. The Camp Fire is the costliest and deadliest wildfire, destroying more than 18,500 buildings. California also endured its largest wildfire on record: the Medincino Complex Fire, burning over 450,000 acres. Additionally, California was impacted by other destructive wildfires: the Carr Fire in Northern California and the Woolsey Fire in Southern California. The total 2018 wildfire costs in California (with minor costs in other Western states) approach $24 billion – a new US record. In total, over 8.7 million acres has burned across the US during 2018, which is

244 / Drought, Flood, Fire

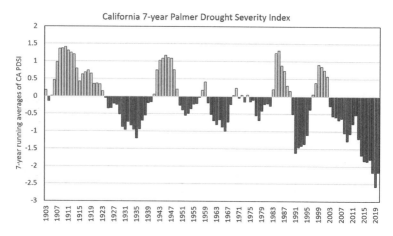

Figure 11.8 California's seven-year average Palmer Drougth Severity Index values. Data from March-to-February data used to allow for an update through January 2020.

well above the ten-year average (2009–2018) of 6.8 million acres. The last two years of US wildfire damage has been unprecedented, with losses exceeding $40 billion.

As with all natural disasters, human systems and human decisions play an important role in setting the stage for crises. In recent years, Californians have built more homes in high fire risk regions. These homes have been connected by a fragile system of electric wires, prone to sparking fires in high winds. Teasing out the complex contributions of human choice, natural climate variability, and climate change is beyond our scope and my expertise. But we can look at low-frequency statewide average Palmer Drought Severity Index values (Figure 11.8) to get a sense of how the balance between moisture supply and atmospheric moisture demand may be changing. These values are based on a March–February annual average to allow for an early 2020 update. A seven-year averaging period was chosen to emphasize the persistent recent dry and warm conditions – the dry conditions leading up to the exceptional 2017 and 2018 fire years. According to the PDSI estimates, the statewide seven-year drought was the most severe on record, and these dry conditions persist. California precipitation pattern is notoriously complex, but the progression toward warmer temperatures and greater aridity seems very likely.

> My friend Gary Eilerts mother lived in Paradise, California. Gary kindly provided the following testimonial. The Camp Fire was incredibly destructive, destroying more than 18,000 buildings and killing 86 people.
> "It took two angels to get me out of Paradise"
> ... said Laura Eilerts.
> She was speaking late in the day on November 8, 2018, the day of her evacuation from the Paradise due to the Camp Fire. On that day, Grandma Laura, as we know her, lost her house, all her belongings, a lifetime of her own paintings, some of her neighbors, her community of friends, her church, and the entire town she'd lived in for 18 years.
> Laura is 91 years-old. She has a lung condition which requires her to be on heavy oxygen 24/7. She can't feel her feet, uses a walker to move about very slowly, and loses her breath very easily. She's been told recently that she doesn't have too long left to live, so she's been meticulously planning every step on that road. The family was told to come and get anything they want before January, and she had planned out everything that needed to happen next, and even after that, so that she could remain where she is happy, in her home in Paradise.
> She's very practiced with fire evacuations from her Paradise home, having experienced more than a few. She has good friends, the Lash family, twenty minutes away, down the ridge into the Central Valley of California, in Chico, who have sheltered her several times when fires have menaced. At her home, she is constantly alert at a very visceral level to the dangers that brushfires present in her normally-dry city in the foothills of the Sierra Nevada mountains, and all it takes is a simple outdoor barbecue in her neighborhood to raise her antennae, focus her attention, and take her away from whatever she was doing at the time.
> So, on a warm and dry Wednesday, November 7th, when the news gave warnings about winds coming the next day, she began mentally preparing for just another evacuation. And on a bright Thursday morning, November 8th, somewhere between seven and

eight in the morning, when the first news came of a brushfire north-east of Paradise, she began acting out a well-practiced routine.

Over a very compressed period of a few minutes, she received a call from the Lashes in Chico, who told her they were immediately sending their son to evacuate her and would drive her back home when it was all over. Shortly afterwards, paramedics arrived at the door ready to evacuate her. But she turned them away, saying "I have friends coming up to take me away; go help others who need it more". Not too long after that, she found out that her friends' son was unable to come and get her, as there were some roads already blocked due to the fires, and others were swamped by evacuating residents. She decided to call the paramedics back, but they told her that it would take a while to come and get her; there were now just too many on the list for pick-ups before her. Still assuming this was still just another evacuation, just like all the others beforehand, her concern rose immediately when she got the clipped official reverse emergency call that basically and starkly told her to: "go now, as quickly as possible".

Standing in her bedroom, still in her pajamas, she remembers briefly considering what she should take with her, aware that each additional step would take time and energy she had little of. She noted her wedding rings in a small heart-shaped ceramic bowl on a bedside table. "I've left them here before", she thought, "and they'll be alright". She grabbed sweat pants and a top as she moved as fast as she could toward the garage, where she kept a car for friends who would drive her to doctor visits. It was now the only sure way to get out, and she had no reservations about doing what she needed to do.

As she opened the garage door, she saw her neighbor, Dan, who lives across the street. He was just getting into his truck with what belongings he could carry, ready to bolt. When he saw the garage door come up, he ran across the street to see if Laura needed help. She told him "no, I'm going to drive myself to Chico", but did ask him to put her heavy oxygen concentrator in the car so she wouldn't have to use oxygen tanks while she waited out the fire in Chico. He did, and then left shortly after her.

Laura got into the car she hadn't driven - because of her feet - and started driving for the first time in many months. "I hope

I don't kill anyone with my driving", she remembers thinking as she concentrated on getting her feet to touch the right pedals.

Her car held her wheelchair, her walker, and the sweat pants and top she grabbed as she left the house. In the back seat, and in the trunk, were several more oxygen bottles, enough to make the car into a pretty substantial moving bomb.

Turning right out of the estate she lived in, onto Sawmill Road, she noted heavy smoke was quickly building up, and already making it difficult to see. A few hundred feet later, she turned left onto what she thought was the main road going West, Pearson Road, but then found it was just a neighborhood road, and followed it back to Sawmill. She got back on Sawmill and went only a few feet more before seeing that flames had just reached the house on the corner of Sawmill and Pearson, and it started becoming clear to her for the first time that this might be more than just an evacuation.

She turned left onto Pearson, merging into hundreds of other cars doing what she was doing, heading towards the main road, the Skyway, out of Paradise. She became a small part of a surging, slow-moving, un-organized caravan of evacuees. But the quick-moving fire was already beginning to out-pace them, and they found they were already having to drive around, under and even through blowing embers, and fires burning in houses and trees along Pearson Road. She began seeing and hearing the sound of what eventually were thousands of propane tanks exploding in houses alongside the caravan, which continued throughout the entire morning and afternoon, giving the evacuation from Paradise a war-like feel.

Not too much farther ahead on Pearson, the caravan had to stop, as the road had become engulfed in fire. From that point on, Laura does not remember which way they turned, where they went, nor how many routes they probed, and does not know at all where she was. She was just "behind the car in front of me", with the fire, and embers, and smoke being much more visible than the road. She does remember that at least three times the snaking caravan ran upon a road entirely blocked by fire and had to stop and turn completely around on a two-lane road. This was a huge logistical maneuver for hundreds of drivers under great duress. But she recalls the unnatural politeness and calm that every driver showed during those moments, slowing and pausing to let others in, giving way to others when needed, and gradually re-starting the desperate caravan's search for a way out.

After several hours of crawling towards a possible exit, only to be blocked by fire, the oxygen tank that Laura was using in the passenger-side front-seat began to run out. She realized she was going to have to hook up a new bottle in the back seat, but it wasn't easily accessible from the front seat. She started struggling to get out of her seat enough to reach it, but it was quickly apparent to her that she wasn't going to be able to reach it, and wouldn't be able to get out of the car.

But soon, the scope of the town's devastation became apparent. The Camp Fire burned more than 18,793 structures, among which 13,972 residences belonging to at least 25,000 people, and killed 86 people in their cars, on the road, and in their homes, the single most devastating fire, ever, in California.

At that very moment, a woman with curly hair, an evacuee like her, in a car moving right alongside of hers must have noticed something, because she decided to stop in the middle of the road and ran over to ask Laura if she needed help. The woman was obviously practiced in transferring breathing devices from one tank to another, and did it quickly, with little comment. Saying

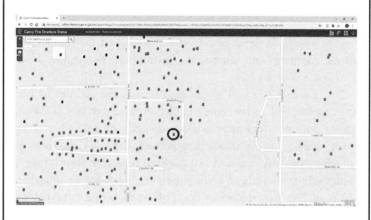

Figure 11.9 Homes destroyed in the Sawmill Road area of Paradise (Grandma's home circled; Red = 100% destroyed). Image credit: Gary Eilerts

nothing more, she finished the job, got back into her car and drove off into the smoke. Laura is positive that unknown woman saved her life right there - a small act of love and concern in the middle of chaos, from a stranger.

It took several more hours before the evacuees that Laura was moving with were able to find a backroad they could use to get through the fire. She finally arrived in Chico - which is a 20-minute drive on a normal day - more than five hours after she left her home. Driving up in front of the store operated by her friends, the Lashes, she was completely exhausted, and completely unable to get out of the car. She honked once, and then waited until they came out to investigate, found her, and helped her out of the car. She ended up staying several days at their home, warmly cared-for, by, as she calls them, "saints".

Grandma Laura, like thousands of her neighbors who began the day as evacuees, ended it as homeless refugees.

A week later, Grandma moved temporarily to her son Gregg's house in San Jose. In the weeks since, she has found a new place to live, not too far from his home, and is ready to create a new life.

Just a few days ago, Gregg went up to Paradise, now accessible after a month of emergency operations, to see if anything is retrievable from the ashes of Grandma Laura's house. Rains that came soon after the fire compacted the ashes and left a sometimes

Figure 11.10 What's left of 5430 Sawmill Road, #14.
Image credit: Gary Eilerts

Figure 11.11 Dan's house was across the street, upper right. The truck in the driveway, upper left, has an "X" to indicate it was checked for bodies. Image credit: Gary Eilerts

toxic mire to dig through. The only things that he found intact were ceramic in nature, born from heat and fire, and able to resist it, like dishes and a few other pieces of pottery. Among these, Laura was surprised to discover that there were two intact ceramic angels, which she doesn't remember ever owning, but which she immediately accepted as being related somehow to the two angels who helped her get out of Paradise.

Gregg also found the heart-shaped bowl. It had only a crumbly residue in it, but nothing that resembled the rings that were last sitting there. Doing a little research, Gregg found that diamonds, unlike many other substances, don't exactly burn – but they do evaporate in high heat, kind of like the entire town of Paradise.

We're all very thankful that Grandma Laura got out of Paradise relatively safely. Many of her friends and former neighbors have been left in very precarious conditions, with no clear way forward from here. But it is still very sad to all of us that the simple pleasure she wanted most - to live out the rest of her time where she belonged, and was known and loved - will not happen.

Figure 11.12 Grandma Laura's two angels.
Image credit: Gary Eilerts

Life is sometimes what happens after you think you have it figured out. But may we all have the grace and courage that Grandma Laura has already shown, while she, and we, continue to figure it out.

12 FIRE AND AUSTRALIA'S BLACK SUMMER

Filled with kangaroos, cockatoos, echidnas, and koalas, Kangaroo Island rests across the Investigator Strait, about 100 miles (155 km), from Adelaide, Australia. Ecologists refer to the Island as "Noah's Ark" due to its rich biodiversity.[1] But in late 2019, half the island burned as a series of vicious fires carved a path of destruction with unprecedented speed and ferocity. The impact on the island's ranchers, farmers, and wildlife was catastrophic.[2] One-third of its 50,000 koalas (approximately 17,000 animals) perished. Thirty to forty percent of its kangaroos died. Many other endangered or rare species experienced severe losses. Species such as the black cockatoo, Rosenberg's goanna, the dunnart, and the short-beak echidna suffered terribly. Some 44,000 animals in total are believed to have perished, and nonprofit groups like the Royal Society for the Prevention of Cruelty to Animals (RSPCA) worked around the clock to save thousands of fire-damaged animals[3] (Figure 12.1).

Tragically, Kangaroo Island's losses were representative of conditions across much of the *entire continent* of Australia.

---

[1] www.bbc.com/news/world-australia-51102658.
[2] www.nytimes.com/2020/02/04/world/australia/kangaroo-island-fire.html.
[3] www.upi.com/Top_News/World-News/2020/01/17/Almost-44000-animals-on-Australias-Kangaroo-Island-have-died-from-fire/5351579283755/.

253 / Fire and Australia's Black Summer

Figure 12.1 A koala in the midst of a devastated landscape on Kangaroo Island. Photo courtesy of the Royal Society for the Prevention of Cruelty to Animals. www.rspcasa.org.au/ki-bushfires-plan/. The RSPCA South Australia provided veterinary support to Kangaroo Island Wildlife Park, and also ran special feeding programs to prevent native animals from starving due to the widespread destruction of their natural habitats and food sources.

Extremely hot, dry conditions – combined with several years of reduced rainfall and a severe 2019 drought (which was related to the Indian Ocean Dipole event discussed in Chapter 8) – set the scene for Australia's unprecedented 2019–2020 fire season.[4]

While many factors contributed to Australia's "Black Summer," in this chapter we will focus on how extremely warm temperatures and associated increases in atmospheric water demand helped set the stage for such a catastrophe. During and before the fire-filled Black Summer, month after month of very warm air temperatures helped increase the atmosphere's ability to draw moisture away from dead and live vegetation, setting the stage for potentially explosive fire expansion. We can start our exploration of these conditions by plotting a time series of annual averages of monthly maximum air termpatures (Figure 12.2). The final bar in this bar plot identifies an exceptionally warm year – the warmest year in the observational record, with an anomaly of +2.4°C when compared to the 1910–1939 mean. The positive anomaly in 2019 is much larger than any previous value back to 1910, when the historical record begins.

Looking at Figure 12.2, we should note a phenomenon we have seen time and time again throughout this book. The past five or so years have been much warmer than conditions during the 1990s, indicating rapid climate change. In this case, recent summers have been about 1°C warmer than the 1990s. Exceptionally warm conditions in 2019 were also preceded by two very warm years in 2017 and 2018. These very warm air temperatures combined with exceptionally low rainfall in 2018. In many areas, the twenty-four-month 2018–2019 rainfall totals were the driest on record. This tendency toward warmer, drier conditions resulted in very dry fuel conditions and extreme fire-risk weather, especially in eastern and southern Australia.[5]

---

[4] Hughes, Lesley, Will Steffen, Greg Mullins, Annika Dean, Ella Weisbrot, Martin Rice; Australia Climate Council. "Summer of Crisis," March 11, 2020. www.climatecouncil.org.au/resources/summer-of-crisis/.

[5] www.bom.gov.au/state-of-the-climate/State-of-the-Climate-2018.pdf.

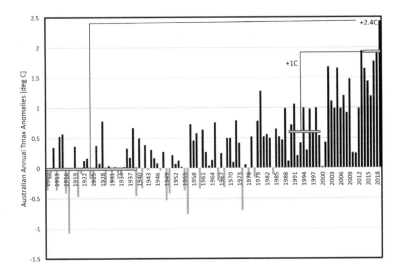

Figure 12.2 Annual maximum air temperature anomalies for Australia. Data obtained from the Australian Bureau of Meteorology. Between 1990–1999 and 2015–2019, maximum air temperatures have increased by approximately 1°C.

These conditions set the stage for massive conflagration. Between September 2019 and early 2020, a stunning 5.8 million hectares of broadleaf forest burned in two of Australia's southeastern provinces, New South Wales and Victoria.[6] Across the continent,[7] Australia's fires burned 72,000 square miles (186,000 square kilometers), destroying some 5,900 buildings and killing more than one billion animals.[8]

An early 2020 Nature Climate Change article set these fires in historic context. Satellite-based estimates of burned area[9] and weather data[10] were used to ask and answer two questions:

---

[6] Boer, Matthias M., Víctor Resco de Dios, and Ross A. Bradstock. "Unprecedented burn area of Australian mega forest fires." *Nature Climate Change* 10.3 (2020): 171–172.
[7] en.wikipedia.org/wiki/2019%E2%80%9320_Australian_bushfire_season.
[8] www.sydney.edu.au/news-opinion/news/2020/01/08/australian-bushfires-more-than-one-billion-animals-impacted.html.
[9] Giglio, Louis, et al. "The Collection 6 MODIS burned area mapping algorithm and product." *Remote Sensing of Environment* 217 (2018): 72–85.
[10] SILO – Australian Climate Data from 1889 to Yesterday, www.longpaddock.qld.gov.au/silo/.

Are the 2019–2020 forest fires unprecedented in scale, and are they the result of unparalleled fuel conditions?

To address the first question, the authors compared the frequency of global fire extent in temperate broadleaf forests over the 2000–2019 time period. This was the period in which satellite-based fire extent estimates were available. The study's results were stunning: "[T]he 2019/20 forest fires have burned a globally unprecedented percentage of any continental forest biome: 21% of the Australian forests." The typical burned area for Australia in previous years was only about 5 percent.

To answer the second question, the authors analyzed the frequency and extent of forested areas in New South Wales and Victoria that were experiencing critically dry "dead fuel moisture" conditions. Dead fuel moisture refers to the fraction of the fraction of the mass of dead forest branches and litter that is made up of water. Forests with low fuel moisture fractions are much more flammable, and under the right weather conditions this allows fires to spread much further and faster. Dead dry fuel conditions can be estimated using daily vapor pressure deficits (VPD).[11] Vapor pressure deficits are based on the difference of two terms, the saturation vapor pressure and vapor pressure. "Pressure," as it is used here, refers to the pressure exerted by an amount of water vapor. In other words, it represents a quantity of water. Saturation vapor pressure defines the theoretical upper limit on the amount of water vapor the atmosphere can hold. It represents conditions when the relative humidity is at 100 percent. As discussed in Chapters 6 and 11, a warming atmosphere can hold more moisture (Figure 6.3). This increased water-holding capacity will lead to increasing vapor pressure deficits, which are tied to increases in forest fire extent (Figure 11.2).

Saturation vapor pressure increases with increasing air temperatures, since there is more room in warmer air to hold water (Figure 6.3). Vapor pressure is simply the actual amount

---

[11] Nolan, Rachael H., et al. "Large-scale, dynamic transformations in fuel moisture drive wildfire activity across southeastern Australia." *Geophysical Research Letters* 43.9 (2016): 4229–4238.

of water vapor in the air. Vapor pressures increase after it has rained, or when moist air blows in from an ocean. The combination of high saturation vapor pressures and low actual vapor pressures produces high vapor pressure deficits. High vapor pressure deficits help desiccate living and dead plants, producing conditions conducive to extensive mega-fires.

The occurrence of large wildfires can be conceptualized as the simultaneous triggering of three switches: (1) the availability of dry biomass to burn, (2) weather conditions conducive to spreading fire, and (3) a source of ignition(s).[12] In southeastern Australia, scientists have documented a two-stage sequence relating vapor pressure deficits to fire. First, atmospheric conditions: vapor pressure deficits are linked to dead vegetation fuel moisture. Dead fuel moisture content refers to the fraction of the mass of the dead vegetation made up by water. Second, the amount of dead fuel moisture is related to the observed extent of wild fires. Figure 12.3 shows the empirical results from a study in southeastern Australia. The gray dots in the background correspond to observed fire extents. The vertical gray bars denote break points at about 10 and 15 percent. When dead fuel moisture levels reach these critical thresholds, the magnitudes of forest fires jump upward.

It is very important that we all understand these results. When dead fuel moisture levels drop below 15 or 10 percent, the risk of a large forest fire increases *dramatically*. This relationship provides extremely useful information for firefighters and first responders. It also helps us understand how high temperatures can set the stage for potentially catastrophic fires. In Australia's eucalyptus forests, fires primarily spread through low-lying forest litter[13]; moist litter acts to limit the spread of fires. Typically, naturally occurring gullies, south-facing slopes,

---

[12] Bradstock, R. A. "A biogeographic model of fire regimes in Australia: Current and future implications." *Global Ecology and Biogeography* 19.2 (2010): 145–158.
[13] Murphy, Brett P., et al. "Fire regimes of Australia: A pyrogeographic model system." *Journal of Biogeography* 40.6 (2013): 1048–1058.

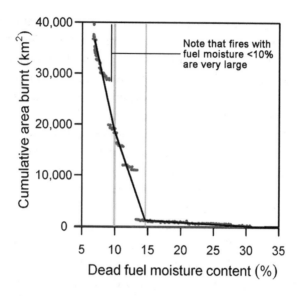

Figure 12.3 Observed dead fuel moisture content from Nolan et al. (2016). Used with permission. The annotation "Note that fires with fuel moisture <10% are very large" was added to facilitate interpretation.

and swamps provide fire-breaks, but under very dry conditions fire extent can grow in a highly nonlinear fashion.

With this background, we can now look at how Boer, Resco de Dios, and Bradstock answered their second question: *Are the 2019–2020 forest fires the result of unparalleled fuel conditions?* Their solution was based on calculations of daily dry fuel moisture values using weather data. In an approach quite similar to our analysis of heat waves (Chapter 5), the authors identified very dry fuel days as whenever a location had a daily dead fuel moisture value of less than 10 percent. This corresponds to the 10 percent break point in Figure 12.3 that is strongly associated with large fires. Calculating the number of days multiplied by the area in forested regions in New South Wales and Victoria produced a metric that helps summarize annual fuel moisture conditions (Figure 12.4). This figure demonstrates that the 2019 fuel moisture levels were truly exceptional, far higher than any on record. We can also see a

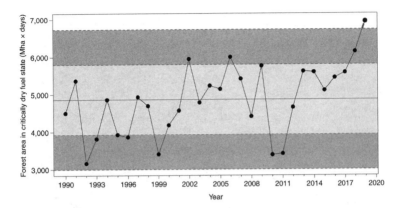

Figure 12.4 Forest area in critically dry fuel state, eastern Australia (1990–2019). Annual variation in the duration and cumulative area of large forest patches (>100,000 ha) in a critically dry fuel state. The horizontal black line indicates the thirty-year mean value; light and dark gray bands indicate mean value ±1 and 2 standard deviations, respectively. Figure and caption taken with permission from Boer, Resco de Dios, and Bradstock (2020)

strong upward trend in the data, a propensity for more forest area to be in a critically dry moisture state. So, yes, *unparalleled fuel moisture conditions helped produce the exceptional 2019–2020 forest fires*.

According to an early 2020 Australian Climate Council *Summer of Crisis* report, these fires impacted 80 percent of Australians[14] and released between 650 million and 1.2 billion tons of CO_2 into the atmosphere, far more than the *annual* emissions of Australia or Germany.[15] In New South Wales, the hardest-hit province, more than 11,000 fires burned 5.4 million hectares (13.3 million acres). The *Summer of Crisis*

---

[14] Biddle, N., B. Edwards, D. Herz, and T. Makkai. "Exposure and the impact on attitudes of the 2019-20 Australian Bushfires." ANU Centre for Social Research Methods (2020). Accessed at: https://csrm.cass.anu.edu.au/research/publications/exposure-and-impact-attitudes-2019-20-australian-bushfires-0.

[15] Bloomberg. "Australia's fires likely emitted as much carbon as all planes" (2020). https://www.bloomberg.com/news/articles/2020-01-21/australia-wildfires-cause-greenhouse-gas-emissions-to-double.

Figure 12.5 The Gospers Mountain Fire rages (December 21, 2019), devastating areas such as Bilpin. From *Summer of Crisis* by Lesley Hughes, Will Steffen, Greg Mullins, Annika Dean, Ella Weisbrot, and Martin Rice.

report detailed the incredible ecological destruction associated with the Black Summer. More than 510,000 hectares (1,260,000 acres) burned in the Gospers Mountain fire near Sydney (Figure 12.5), making it the largest forest fire recorded in Australia. Fires also burned through *80 percent* of the Blue Mountains World Heritage Area, and through more than half of the Gondwana Rainforests in New South Wales and Queensland. These ancient forests include unique species that date back to the time of the Gondwana supercontinent, some 180 million years ago. For millions of years, too damp to burn, the Gondwanan rainforest evolved in a fire-free state. In 2020, according to initial assessment, more than 300 species, many of them endangered, have had at least 10 percent of their habitat burned by the fires.

We can use climate change simulations to place the warm New South Wales and Victoria air conditions in an historic context. Figure 12.6 shows average annual southeastern Australia air temperature anomalies for a large collection of

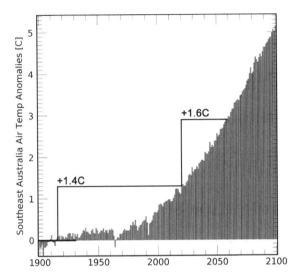

Figure 12.6 Average New South Wales/Victoria air temperature anomalies from a collection of climate change simulations.

eighty-one simulations.[16] Between the late nineteenth century and early twenty-first century, the models estimate an average increase of about 1.4°C. If we project out into the future using a rapid emission scenario,[17] the models anticipate that this area of southeastern Australia will continue to warm rapidly. By 2060, we appear likely to see another approximately 1.6°C increase in *average* temperatures. Natural year-to-year variations in temperature will result in years even warmer than this average.

We can use this large collection of climate change simulations to perform a formal climate attribution analysis. This allows us to assess how climate change contributed to the exceptionally warm conditions in southeast Australia. We use a large set of simulated 1900–1929 air temperatures to describe natural year-to-year variations in a world without climate change. The

---

[16] Based on an eighty-one-member ensemble of climate change simulations from phase 5 of the Climate Change Model Intercomparison Project (CMIP5). Simulations obtained from the KNMI climate explorer.
[17] The 8.5 W/m² Representative Concentration Pathway scenario. Discussed further in Chapter 13.

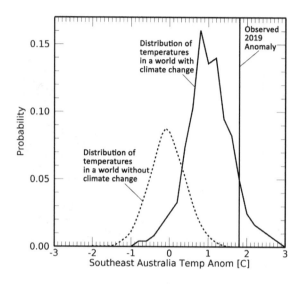

Figure 12.7 Formal attribution analysis of New South Wales/Victoria air temperature anomalies from a collection of climate change simulations.

distribution of these temperatures, expressed as differences from the 1900–1929 average, is shown with the dashed line in Figure 12.7. Typical values range from about −1°C to +1°C. The observed 2019 temperature anomaly of +1.8°C is way warmer, allowing us to formally state that *temperatures as warm as those observed in southeastern Australia in 2019 would have been essentially impossible in a world without climate change.*

We can also use this large collection of climate change simulations to examine how likely a year like 2019 might be in a world *with* climate change. The solid line in Figure 12.7 shows the distribution for 2019 based on the climate change simulations forced with observed changes in greenhouse gasses and aerosols. While the +1.8°C lies on the upper side of this distribution, it is quite likely to have conditions this warm, according to the models, in a world with climate change. So, according these models, *a year as warm as 2019 should happen quite frequently, about once every nine years, in a world with observed concentrations of greenhouse gasses and aerosols.*

The strong link between vapor pressure deficits and fire extent (Figure 12.3) allows us to connect these increases in air temperatures directly to changes in dead fuel moisture conditions (Figure 12.4). We can do this by performing an attribution analysis similar to that performed in a 2019 climate attribution study focused on the 2018 Four Corners drought in the United States.[18] Because there is a direct relationship between increases in air temperatures, saturation vapor pressure, and dead fuel moisture levels, we can translate changes in temperature to changes in dead fuel moisture.

Using the same observed air temperature and vapor pressure data as analyzed by Boer, Resco de Dios, and Bradstock, we can use a little simple mathematics to calculate time series of annual vapor pressure deficits and saturation vapor pressure values. In Figure 12.8, these values are averaged across New South Wales and Victoria, and expressed as anomalies from the 1990–2019 mean. The black bars in Figure 12.8 show the observed annual vapor pressure deficits (saturation vapor pressure minus actual vapor pressure). There is a very strong upward trend (shown with a diagonal line). Note that the 2020 vapor pressure deficit was by far the largest on record – about twice as large as any value prior to 2018.

The straight diagonal line in Figure 12.8 displays a regression-based vapor pressure deficit trend estimate. Interannual vapor deficit variations correlate fairly well with this trend, with a correlation value of 0.6. Using this trend line to project out to 2040 presents a scary potential future. *In just twenty years, a persistence of the observed trend would mean that average years in New South Wales might have vapor pressure deficits similar to the worst year on record (2019).*

The vertical gray bars in Figure 12.8 show year-to-year variations in saturation vapor pressure. These variations are

---

[18] Williams, Emily, et al. "Quantifying human-induced temperature impacts on the 2018 United States Four Corners hydrologic and agro-pastoral drought." *Bulletin of the American Meteorological Society* 101.1 (2020): S11–S16. journals.ametsoc.org/doi/pdf/10.1175/BAMS-D-19-0187.1.

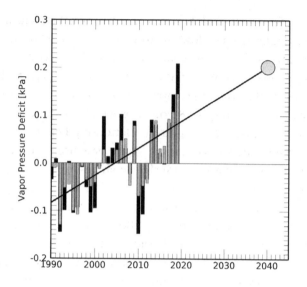

Figure 12.8 Time series of observed annual average vapor pressure deficits (black bars) and saturation vapor pressure values (gray bars) for New South Wales and Victoria. Also shown is trend estimate for 2040.

driven purely by changes in air temperature. The magnitude of the trend in this data is essentially identical to the trend in vapor pressure deficits, but the correlation is higher (0.7). Warming air temperatures are largely responsible for the upward trend shown in Figure 12.8. It is also very noteworthy that in 2018 and 2019, saturation vapor pressures were exceptionally high as well. These data suggest that in both 2018 and 2019, the impact of warm temperatures was even greater than the impact of below normal atmospheric water vapor. Two back-to-back very warm years help set the stage for fire.

Finally, we can connect the dots, linking our attribution analysis (Figures 12.6 and 12.7) with estimates of vapor pressure deficits and areas with exceptionally dry fuel moisture conditions (Figure 12.9). The x-axis of Figure 12.9 shows annual 1990–2019 vapor pressure deficit values, averaged over New South Wales and Victoria. The y-axis shows estimates of the area-days with dead fuel moisture values of less than 10 percent. As in Figure 12.8, the year 2019 – and, to a lesser

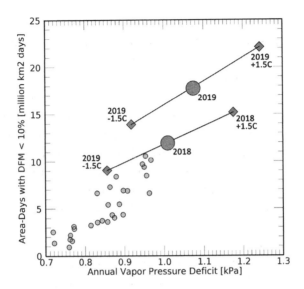

Figure 12.9 Scatterplot of New South Wales/Victoria annual vapor pressure deficit values and area in critically dry fuel state.

degree, 2018 – stand out as exceptional. With an estimated 17 million km²-day area-days in critically dry fuel state, the observed exposure levels in 2019 were about 70 percent above the previous maximum values of about 10 million.

Our climate attribution analysis (Figure 12.6) suggested that about 1.5°C of the 2019 warming can be associated with long-term changes in the mean state due to climate change. This allows us to produce a "counterfactual" experiment that describes a world without this warming. To simulate a world with vapor pressure deficits similar to 2019, but with cooler conditions similar to a world without climate change, we can cool our observed weather data by 1.5°C and recalculate the daily pressure deficits for each location. Then, for each location, we can recalculate the dead fuel moisture levels. Then we can calculate the associated area with critically dry fuel conditions. These results are shown with the left-hand diamonds in Figure 12.9. Even with 1.5°C of warming removed, the year 2019 would still have had exceptionally dry fuel conditions. Naturally occurring variations in rainfall and air temperatures were likely the primary driver of these arid conditions. Nevertheless, a substantial fraction of the critically dry area-days (~30 million km²-days) appears related to the human-induced warming of 1.5°C.

We can also use the projected near-term warming (Figure 12.6) to estimate that southeastern Australia may soon be, on average, about 1.5°C warmer than at present. To explore what a year like 2019 might look like, we can add 1.5°C to the observed weather data and recalculate the 2019 statistics. These results are shown with the right-hand diamonds in Figure 12.9) An exceptionally warm/dry year like 2019 may soon appear much worse, from a fire weather perspective. Our future climate projection indicates a massive increase of about 5 million km²-days in areas with critically dry fuel conditions. Future dry fuel events worse than 2019 may be on the way soon.

Conclusion

Mega-fires are very complex, and arise through an interplay of ignition, weather conducive to fire, and very dry fuel conditions. The simplest aspect of this complex causal network is the fairly straightforward relationship between extremely dry fuel conditions and wildfire extent. When fuel conditions are extremely dry, natural barriers to fire spread are greatly diminished. Given that there is a straightforward relationship between climate change and increasing annual temperatures (Figures 12.2, 12.6, and 12.7) and increases in saturation vapor pressure-driven decreases in dead fuel moisture, we have analyzed annual vapor pressure values to gain insights into one of the key drivers of the 2019–2020 fire season in New South Wales and Victoria.

One simple but important takeaway from this analysis is that air temperature-related impacts are extremely important for fire hazards. Rainfall totals have been very low in Australia, and we might have found that vapor pressure deficits were the main driver of the total vapor pressure deficit anomalies. If that was the case, then we would have found large black bars and small gray bars in Figure 12.8. Examples of such vapor-pressure deficit-driven years can be seen in 2002 and 2006. By contrast, in 2018 and 2019 we see that the gray bars themselves are large. In fact, in 2019, the results presented here suggest that the very warm temperatures, acting alone, would have produced vapor pressure deficits greater than any previous year since 1990. These very warm conditions would have been *impossible* in a world without human-induced climate change, and are almost *certain* to become more frequent in the near future.

So, while exceptionally warm temperatures were just one enabling component of the 2019–2020 fire season, they certainly did help produce the exceptionally dry 2019 fuel conditions in New South Wales and Victoria. Climate change contributed in a major way to Australia's exceptionally dangerous and destructive Black Summer by increasing vapor pressure deficits and fuel moisture conditions, allowing fires to spread. Going forward, both observations (Figure 12.8) and climate change simulations (Figure 12.6) indicate that these exceptionally dry conditions are likely to become more frequent, and soon. As discussed in the next chapter, the carbon dioxide emitted from these types of massive fires may act as a positive feedback, contributing to more global warming and a destabilization of our "Dixie® Cup" planet.

13 DRIVING TOWARD +4°C ON A DIXIE® CUP PLANET

Representative Concentration Pathways

To explore our collective future, scientists develop plausible emission, pollution, and land use scenarios. Some of these scenarios describe rapid economic and population growth, and concomitant increases in fossil fuel emissions. Other scenarios describe a rapid transition to clean energy. In keeping with our standard desire to baffle and obfuscate, us scientists call these "scenarios" representative concentration pathways (RCPs). "Scenarios" just seems so transparent and blasé. Four main RCPs[1] are used to project out to 2100, providing plausible scenarios (what-if stories) that we can use to explore the potential greenhouse impacts of different human development pathways. Do we face a cheery green future, a dark rapid emissions outlook, or something in between? The RCPs help us understand how future changes in population, gross domestic product (GDP), energy use, land use, mitigation,

---

[1] Vuuren et al. "The representative concentration pathways: an overview." *Climatic Change* 109 (2011): 5–31, DOI 10.1007/s10584-011-0148-z. The latest work featured by the Intergovernmental Panel on Climate Change uses "shared socioeconomic pathways," which provide a more complete description of potential future social configurations. Here we focus on the older, simpler "representative concentration pathway" framework because it is easier to explain and still widely relevant.

and pollution might change the atmospheric forcing associated with greenhouse warming.

"Concentration" refers to estimates of the amount of different greenhouse gasses and aerosols in the atmosphere at different times. These estimates are a function of emissions, land use, and mitigation activities. The RCP2.6 scenario[2] is a happy story characterized by a near-term peak in emissions, followed by a rapid decline in emissions. At the other end of the spectrum, RCP8.5[3] is characterized by rapid population growth, limited technological innovation, and rapidly increasing greenhouse gasses. The slightly less grim RCP6[4] scenario is characterized by increasing greenhouse gas emissions over time, but with stabilization shortly after 2100 due to the application of a range of technologies and strategies for reducing emissions. The RCP4.5[5] scenario has relatively slow increases in greenhouse gas emissions and the stabilization of greenhouse gas emissions shortly after 2100. These scenarios also represent different outcomes in terms of "energy intensity" and "carbon fraction." Energy intensity, the energy per unit of income, refers to how fuel efficient an economy is. Technology, investment, energy efficiency, and transitions toward higher-value products and service industries make some economies (like Germany's) less energy intense. Carbon fraction refers to the emissions per unit of energy used in an economy; moving toward renewable energy reduces an economy's carbon fraction. RCP6 and RCP8.5 have

---

[2] Riahi, K., A. Grübler, and N. Nakicenovic. "Scenarios of long-term socio-economic and environmental development under climate stabilization." *Technological Forecasting and Societal Change* 74 (2007): 887–935.
[3] Van Vuuren, D. P., M. G. J. Den Elzen, P. L. Lucas, B. Eickhout, B. J. Strengers, B. et al. "Stabilizing greenhouse gas concentrations at low levels: An assessment of reduction strategies and costs." *Climate Change* 81 (2007): 119–159.
[4] Clarke, L. E., J. A. Edmonds, H. D. Jacoby, H. Pitcher, J. M. Reilly, and R. Richels. "Scenarios of greenhouse gas emissions and atmospheric concentrations. Sub-report 2.1a of Synthesis and Assessment Product 2.1." Climate Change Science Program and the Subcommittee on Global Change Research, Washington, DC, 2007.
[5] Fujino, J., R. Nair, M. Kainuma, T. Masui, and Y. Matsuoka "Multigas mitigation analysis on stabilization scenarios using aim global model." *The Energy Journal Special Issue* 3 (2006): 343–354.

high carbon factors throughout the twenty-first century. RCP8.5 maintains a high level of energy intensity.

RCP2.6, RCP4.5, RCP6, and RCP8.5 indicate the anticipated 2100 radiative forcing value in Watts per meter squared [Wm^{-2}]. The "radiative forcing value" is the change in the amount of long-wave radiation absorbed by Earth's surface due to greenhouse gas and aerosol emissions. We are now at a radiative forcing of about ~2 Wm^{-2}. The future radiative forcing values range from small changes (2.6 Wm^{-2}, or ~130% of the current radiative forcing value of 2 Wm^{-2}) to very high (8.5 Wm^{-2}, or about ~425% of the current value). Hence the numbers after the RCPs.

In terms of human and environmental impacts, the differences between these scenarios are profound. In RCP2.6, radiative forcing peaks very soon, shortly after 2025, and declines to levels near those of today. In RCP8.5, conversely, rapid CO_2 and methane emissions quickly lead to runaway increases in radiative forcing. By 2050, the RCP8.5 radiative forcing has already more than doubled over the 2000 value, to about 5 Wm^{-2}. In RCP4.5 and RCP6, on the other hand, mitigation and reduced emissions eventually lead to a stabilization in greenhouse gas levels. This occurs around 2070 for RCP4.5 and after 2100 for RCP6. By 2050, these latter two scenarios also predict substantial warming, with increases in radiative forcing of about 3.7 Wm^{-2}.

The differences between these forcing numbers make a big difference in where we are heading (Figure 13.1), even on a short time scale like the change between now and 2050. The RCP4.5 scenario suggests an additional about 1-degree warming between 2000 and 2050, and about 2-degree warming by the end of twenty-first century. The RCP8.5 scenario suggests an additional about 1.5-degree warming between 2000 and 2050, and a greater than 3-degree warming by the end of twenty-first century. This latter scenario would be very, very bad.

In RCP8.5, population approaches 10.5 billion by 2050 and peaks near 12 billion in 2100. In RCP4.5, population

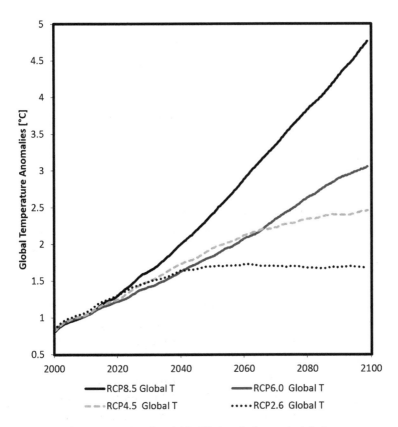

Figure 13.1 Simulated (CMIP5) and observed global surface temperatures.

totals of about 9 billion in 2050, then levels out. RCP4.5 projects that our total 2100 energy consumption will be about twice our current level. In the RCP8.5 scenario, however, rapid population growth and a lower rate of technology growth results in a rapid increase in energy consumption, resulting in 2100 consumption rates of more than *three* times those at the beginning of the twenty-first century. The mix within the total energy consumed varies substantially, as well, with RCP8.5 accounting for most of its increase in energy through oil consumption. RCP2.6 represents a scenario of aggressive energy substitution, and oil consumption begins dropping before 2025. While RCP4.5, RCP6, and RCP2.6 all predict similar levels of energy consumption, the transition toward alternative energy sources limits rapid increases in radiative forcing in RCP4.5 and RCP2.6.

Changes *early* in twenty-first century emissions will have important compounding impacts on the total amount of greenhouse gas in the atmosphere. It is important to note that greenhouse gasses have long residence times in the atmosphere. Carbon dioxide is thought to have a residence time of between 30 and 95 years[6] while nitrous oxide has a mean residence time of 114 years.[7] Methane (CH_4), on the other hand, has a relatively short residence time (14 years). Because greenhouse gasses persist in the atmosphere, they build up over time. This can, and will, if business-as-usual continues, have dire consequences for our planet. Linear 2000–2060 increases in RCP8.5 CO_2 emissions, for example, translate into exponential increases in 2000–2100 CO_2 concentrations. The converse also holds. Were we to adopt a rapid near-term transition away from fossil fuels, the RCP2.6 scenario suggests that we would quickly see the increase in atmospheric CO_2 stop, stabilizing the climate system.

And what if we wait? It depends for how long. Unless we quickly transition to an RCP2.6-like path very soon, the early twenty-first-century accumulation of greenhouse gasses will likely propel us beyond the dangerous 2°C warming threshold by 2050. Besides having devastating impacts on agriculture, such warming would also heavily disturb eco-climatic zones, endangering many species of animals, and heavily impact coral reefs and associated fisheries.[8] Remember that this would be warming at a rate that is unprecedented – as far as we know – in Earth's history: animals and plants simply cannot migrate or adapt – and certainly not evolve – fast enough to keep up with

---

[6] Jacobson, M. Z. "Correction to 'control of fossil-fuel particulate black carbon and organic matter, possibly the most effective method of slowing global warming.'" *Journal of Geophysical Research* 110 (2005): D14105. doi:10.1029/2005JD005888.

[7] IPCC Fourth Assessment Report, Table 2.14, Chap. 2, p. 212.

[8] "Turn down the heat," Worldbank report. https://www.worldbank.org/en/topic/climatechange/publication/turn-down-the-heat

the rate of change. We are already seeing this expressed as catastrophic coral bleaching,[9] massive marine fisheries impacts,[10] and fire-induced animal genocide.[11] But this affects humans too. As we saw in Chapter 5, increased temperatures will expose billions of people to temperatures beyond our body's adaptive capacity. In Africa,[12] a 2°C warming by 2040–2050 is likely to produce substantial declines in crop yields, a 30–40 percent increase in the rate of crop failure (from about once in five years to once in four years), more frequent extreme warm events, 30–50 centimeters of sea level rise around the coasts, and put an additional 10–15 percent (above now) of sub-Saharan African species at risk of extinction.

In Southeast Asia (Pakistan, India, Nepal, Bhutan, Bangladesh, Sri Lanka) by 2050, some fisheries may experience a 15–50 percent decline in productivity, and 98–100 percent of coral reefs are likely to become "thermally marginal" – which means that the water has gotten so warm, it will be difficult for them to survive. During summer, 60–70 percent of the region will experience unusual heat extremes, and 30–40 percent of the region will likely experience *unprecedented* heat extremes. In Southern Asia, the World Bank finds that a 2°C warming will likely be associated with a 20 percent increase in unusual summertime heat extremes, while a sea level rise might expose millions more people to flood risk, especially in Bangladesh. Melting glaciers may also reduce water availability in the Ganges, Brahmaputra, and Indus watersheds. Without climate change, the number of moderately stunted children is anticipated to decline to 11 percent of the total number of children by 2050; with climate change, the +2°C scenario, the percentage

---

[9] Hughes, Terry P., et al. "Coral reefs in the Anthropocene." *Nature* 546.7656 (2017): 82–90.www.nature.com/articles/nature22901.
[10] Webb, Robert S., and Francisco E. Werner. "Explaining extreme ocean conditions impacting living marine resources." *Bulletin of the American Meteorological Society* 99.1 (2018): S7–S10.
  https://journals.ametsoc.org/doi/pdf/10.1175/BAMS-D-17-0265.1.
[11] www.sydney.edu.au/news-opinion/news/2020/01/08/australian-bushfires-more-than-one-billion-animals-impacted.html.
[12] "Turn down the heat," Table 3.4.

of moderately stunted children only declines to 14 percent. With a 2050 estimated population of 2.2 billion for Southeast Asia, this 3 percent difference might represent some 70 million children more moderately stunted children. Pause and think about that for a second.

Looking further out, however, we can see a huge difference between the responsible and plausible RCP4.5 scenario and the extremely dangerous RCP8.5 scenario (Figure 13.1). So far, we have "only" experienced about 1°C of warming – and in this book we have already explored some of the mayhem related to that warming. A consistent theme of this book has been that warming will not happen in an even-handed way. Some regions, like California, have warmed much faster than the global average (Chapter 11). Half a degree or more of extra warmth in the tropical oceans can amplify the intensity of Indian Ocean Dipole, El Niño and La Niña–related droughts (Chapters 8, 9 and 10). And a warming atmosphere will produce a world with more extreme heat waves and rainfall events (Chapters 5–7) and very dry fuel moisture conditions (Chapters 11 and 12).

Looking forward, we face a collective choice, probably ranging from a total warming of about 2°C to possibly more than 4.5°C. And as we have already seen, climate change has *already* made weather and climate extremes more dangerous.

Most concerning, perhaps, might be the potential for runaway warming. The Earth, like many complex systems, is a delicate yet robust network of energy balances and feedbacks. Yes – delicate *and* robust. The Earth system has negative feedbacks that can help maintain thermal equilibrium. For example, as the concentration of carbon dioxide increases, the oceans and rainforests, up to a point, can absorb more CO_2. Conversely, there are positive feedbacks for climate change, and "positive" in this context is not a good thing. For example, if global warming melts our ice caps, they will reflect less energy, the Earth will absorb more energy, and the rate of global warming will increase. I like to think of these feedbacks as a Dixie® Cup system (Figure 13.2), named after a small inverted cardboard "Dixie®" cup. Dixie® cups are small water glasses that you find

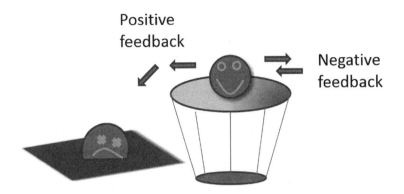

Figure 13.2 Schematic of a Dixie® Cup Planet.

in dentist offices. My conceptual Dixie® Cup system has the Earth rolling around on the top of this inverted cup. Typically, a Dixie® Cup system maintains homeostasis. Small perturbations in temperature are met with small contrary responses, maintaining a temperature equilibrium. The Earth system maintains its position within its basin of attraction – the top of the Dixie® Cup. But huge perturbations might tip the system over the edge.

Yours truly (the author) is a Dixie® Cup system. If Planet Funk is hot, Planet Funk sweats. If Planet Funk is cold, Planet Funk shivers.

Yet Planet Funk might be pushed beyond operating norms. A runaway fever might kill invading bacteria but bake the Funk brain along the way. Funk too cold might lay down in the snow and sleep forever.

So too Planet Earth sits within a similar "basin of attraction" in system space; small perturbations are met with negative feedbacks, leading to system stability.

On Planet Earth, many negative feedbacks act to naturally sequester atmospheric carbon dioxide. If the Earth warms very slowly, plants' metabolic processes increase, pulling more CO_2 out of the atmosphere. Plants take in atmospheric CO_2 through small holes in their leaves called stomata. Through the miracle of photosynthesis, this CO_2 is combined with water and solar energy to make sugar. Plants, however, face a fundamental trade-off. They have to open their stomata to get CO_2 but in so doing they lose precious H_2O. When there is more CO_2 in the atmosphere, plant photosynthesis is more efficient, and this acts as a negative feedback, assuming there is sufficient water available. There can be lots of other negative feedbacks, with the oceans absorbing more CO_2 as the Earth warms; molds, fungi, and insects breaking down waste products faster; and vegetation in northern regions growing faster and pulling more CO_2 out of the atmosphere. These and other negative feedbacks have helped keep temperatures on our "Goldilocks Planet" in a very narrow range for most of our planet's past.[13]

Within a narrow range of temperatures, these negative feedbacks act to maintain homeostasis, and Planet Ours is fine. Pushed beyond the rim of the cup, however, these negative feedbacks may fail, and destructive positive feedbacks may ensue. We know that things can go pear-shaped rather quickly. At least once in our planet's past, we experienced freezing conditions that geologists refer to as "Snowball Earth." During the distant past, the Earth grew so cold that even the tropical oceans froze, and this acted as a positive feedback, as frozen waters reflected back more of the sun's energy. The opposite process may be already under way, as the polar ice caps shrink precipitously. Disappearing ice and snow no longer reflect downwelling solar radiation back to space. The modeling of sea ice loss is tremendously difficult, and sea ice responses represent one of the great uncertainties in climate change research. For the Arctic, we do have compelling satellite data showing very large declines in sea ice extent (and thickness) since 1979 (Figure 13.3). Between 1979 and 2019, minimum sea ice extents have decreased by almost 39 percent, or 2.73 million km^2.

Warming may also impact the vast tracts of land covered with permafrost. These thawing lands of ice and snow may emit more carbon and methane, amplifying global warming

---

[13] Zalasiewicz, Jan, and Mark Williams. *The Goldilocks Planet: The 4 Billion Year Story of Earth's Climate*. New York: Oxford University Press, 2012.

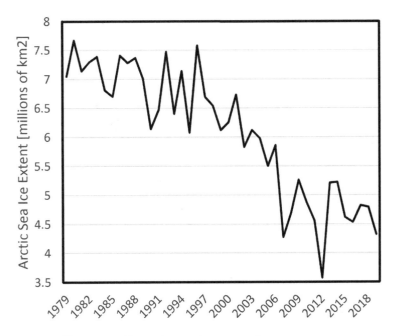

Figure 13.3 September sea ice extents from 1979 to 2020. Data provided by the National Snow and Ice Data Center, accessed February 28, 2020.

in unexpected and nonlinear ways. Increases in global fire-related emissions or reductions in the respiration rates of tropical forests or the sequestration of CO_2 by the oceans could also help push planet Earth into a rapid warming trajectory.

Given that even a 2°C warming by mid-century will have serious implications for food security, crop production, fisheries, and corals, we need to pay close attention to the path we are following. Based on recent data, which seems to be the most representative concentration pathway?

Examining the Global Carbon Budget

Since I'm fantastically and fatalistically addicted to data, I went to the source to answer this question – downloading the newly released 2019 Global Carbon Budget data from the

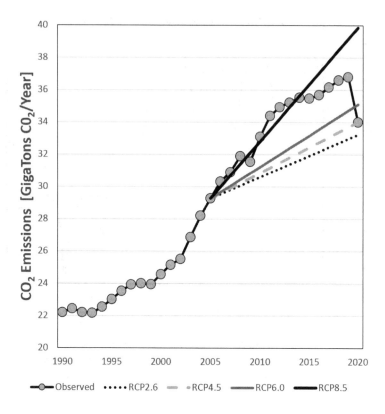

Figure 13.4 A time series of 1990–2020 observed emissions is shown with filled circles. Straight lines depict emission scenarios. Note the observed drop in 2020 emissions was due to impacts from the COVID19 pandemic, as discussed in https://www.globalcarbonproject.org/carbonbudget/20/files/GCP_CarbonBudget_2020.pptx. Reduced economic activity and travel reduced emissions. How these impacts unfold in a post-pandemic world remains unclear.

Global Carbon Project (www.globalcarbonproject.org).[14] This group closely monitors atmospheric greenhouse gas emissions and concentrations and posts their analyses on their website. Here we examine their 2019 carbon budget results. Figure 13.4 shows 1990–2019 observed global fossil fuel and cement

[14] http://www.globalcarbonproject.org/carbonbudget/.

emissions[15] in gigatons of CO_2. At the Earth's standard pressure, this much CO_2 would create a one-meter layer some 2,125 by 2,125 kilometers; an area of 5 million square kilometers covering half the United States.

Figure 13.4 supports both alarm and hope. Between 1990 and 1999, emissions rose at about 1 percent per year. Between 2000 and 2009, emissions increased almost three times faster, by about 3 percent per year, a rate that is slightly faster than the IPCC's dire RCP8.5 scenario. To visualize this rate of change, we can imagine total atmospheric CO_2 as one-meter-thick goo (like John Carpenter's *The Fog*), spread across the eastern half of the United States. By 2035, a continued 3 percent per year increase covers the rest of the country (and Hawaii and Alaska).

But Figure 13.4 also shows signs of hope. Note how since 2015 the increase in global CO_2 emissions has fallen below the dangerous RCP8.5 increases. Reductions in the use of coal and other mitigation efforts appear to be moving us toward the more modest RCP6.0 or RCP4.5 scenarios, scenarios that may ultimately produce 2.5°C to 3.3°C of warming.[16]

We should all note that the near-term differences between RCP8.5 and RCP6.0 or RCP4.5 are huge. Following RCP8.5, we arrive at a ~3°C warming by 2060 (Figure 13.1), enough warming to leave our children in a world of hurt, literally, and well on our way to a dangerous more than 4°C warming by century's end.

As we have already seen in previous chapters, this warming won't just manifest as a smooth, low-frequency increase in ocean and atmospheric temperatures. We will see this warming manifest as more intense weather and climate extremes – stronger heat waves, droughts, and floods, as well as more intense climate anomalies like El Niños and La Niñas.

---

[15] It takes about 0.5–0.8 pounds of CO_2 to produce 1 pound of Portland cement (www.co2list.org). Besides water, concrete is the most commonly used material on earth, and accounts for about 1% of U.S. carbon emissions; http://www.concretethinker.com/technicalbrief/Concrete-Cement-CO2.aspx.

[16] Based on estimates provided by the 2019 Global Carbon Budget, https://www.globalcarbonproject.org/carbonbudget/.

How old will your children, nieces, or nephews be in 2060? My daughter Amelie and son Theo will be sixty-eight. Will they remember us as the generation that failed? If we reach 3 degrees of warming, we can expect a much greater frequency of unusual and unprecedented heat extremes, declines in crop yields, increases in malnutrition and childhood stunting, ocean acidification, and the destruction of many coral reefs. We may also face a real risk of runaway warming, and a dramatic drop of livability on our Dixie® Cup Planet (Figure 13.2) as melting polar ice caps reflect less heat, the oceans acidify, and the carbon uptake of vegetation diminishes.

Who is to blame? What caused the rapid increase in emissions? Are these the same questions? If we look at total cumulative emissions since 1959, North America and the twenty-eight European Union nations have contributed about 55 percent of global emissions, with Asia contributing about 31 percent. From this perspective, the United States and Europe have produced about half of the warming.

Looking just at the recent acceleration, however, we see that the relative contributions to annual emissions have changed dramatically. Between 2000 and 2019, global carbon emissions *increased* by 50 percent. We are literally stepping on the gas pedal (or coal pedal, actually), just as we race head-first for the cliff. Most recent emissions come from Asia, and most of those increases involve increasing coal use. Between 2000 and 2018, global coal emissions *increased* by 62 percent. Since 2010, however, coal emissions have stabilized, and were overtaken by growth in natural gas emissions.

As the developing world develops further, boosting gross domestic products and raising household incomes, the bootstraps they pull themselves up with are coated with the same cloying soot as nineteenth-century England. So, topping our list of 2018 emission contributors is coal, coming in at 40 percent of total emissions. It is only since 2005 that coal has exceeded oil as the biggest emission source. In 2017, the top four emitters covered 60 percent of total emissions: China (27%), United States (15%), the twenty-eight European Union

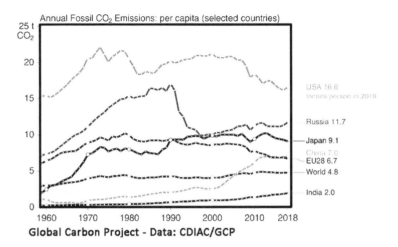

Figure 13.5 Per capita CO_2 emissions from 1959 to 2017. Figure from the Global Carbon Project's "Carbon Budget 2018."

countries (10%), and India (7%). Emissions growth is by far the greatest in Asia, but on a per capita basis, the United States is far and away the largest contributor to global warming (Figure 13.5). The global average per capita CO_2 emissions rate is about 5 tons per person. The United States has historically been at about four times that (20 tons per person). In 2018, the US emission rate was about 16 tons per person, still far more than Europe or China, which come in at around 7.5 tons per person. What if the United States we to cut per capita emissions to European levels? If 335 million people emitted 9 tons of carbon less per year, that would reduce global emissions by about 3 gigatons. That would be a big, big help.

So how bad are we doing? Our current path appears similar to the more rapid emission growth scenarios. Our brief analysis of the change in emissions, placing us on this path, underscored the important role played by emissions increases by China and other emerging economies. As industrialization has swept the developing world, the global distribution of emissions has shifted dramatically as well. In 1990, developed countries accounted for 65 percent of all emissions. In 2011, they accounted for 42 percent. Since 2000, developing countries have

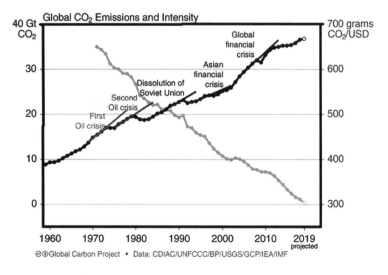

Figure 13.6 Global CO_2 emissions from fossil fuel use and industry since 1959 and emissions intensity CO_2/GDP. From the 2018 Global Carbon Budget analysis. Economic activity is measured in purchasing power parity (PPP) terms in 2010 US dollars.

exhibited a consistent annual increase in emissions of 0.24 PgC per year, while the developed countries emissions have remained steady. While slower growth or reductions in the United States and Europe could help reduce emissions substantially, another key determinant of twenty-first-century global health will be the energy economies of Asia and the southern hemisphere.

The Global Carbon Project has published a recent assessment of emission projections.[17] In this opinion piece they focus on our need to reduce our Emissions Intensity – the amount of CO_2 emissions necessary to produce a certain amount of productivity, i.e., emissions divided by GDP. Global emissions are increasing as emissions intensities decrease (Figure 13.6). Like many global trends, we find opposing tendencies. Our economies are growing, leading to increased emissions, but our economies are also growing more efficient. To reduce emissions while feeding, housing, and employing a

[17] Jackson, R. B., et al. "Reaching peak emissions." *Nature Climate Change* (2015), doi:10.1038/nclimate2892.

rapidly growing global population, we need to rapidly transition away from coal and oil, adopting cleaner and more renewable sources energy and industries.

A recent *Nature Commentary* by some of the Global Carbon Project scientists[18] stresses the need to convert renewable energy sources. Their message is directly in line with the story of this book:

> Every year of rising emissions puts economies and the homes, lives and livelihoods of billions of people at risk. It commits us to the effects of climate change for centuries to come. Already, the terrible impacts of 1°C of warming above pre-industrial levels are evident. Disasters triggered by weather and climate in 2017 cost the global economy US$320 billion, and around 10,000 lives were lost.[19] The full costs of 2018's disasters have yet to be tallied – including Typhoon Mangkhut, hurricanes Florence and Michael, and the heatwaves and wildfires that have ravaged swathes of Europe and the United States. These events are likely to contribute to an exponential rise in damages, amounting to some $2.2 trillion over the past two decades.[20]
>
> When it comes to rises in global average temperature, every fraction of a degree matters. A report published in October by the Intergovernmental Panel on Climate Change (IPCC) projected devastating impacts at 2°C. These include the loss of almost all the world's coral reefs, and extreme, life-threatening heatwaves that could affect more than one-third of the world's population.[21] Limiting warming to 1.5°C will significantly lessen those impacts.

Striking, strident, true. We have experienced a warming of 1°C already, and we are already seeing dangerous increases in climate extremes. Increases in climate extremes that are hurting

---

[18] https://www.nature.com/articles/d41586-018-07585-6.
[19] https://www.munichre.com/topics-online/en/climate-change-and-natural-disasters/natural-disasters/2017-year-in-figures.html.
[20] https://www.unisdr.org/archive/61121.   [21] https://www.ipcc.ch/sr15/.

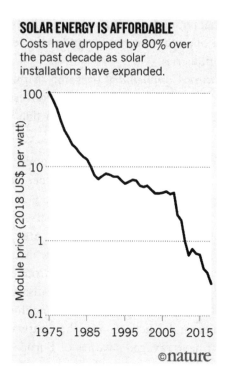

Figure 13.7 Cost for solar power in US dollars per Watt. From Figueres et al. (2018)

people. But this *Nature Commentary* also stresses reasons for hope. Alternative energy is exactly that – a *viable alternative*. Solar power has dropped from a cost of US$100 per Watt in 1975 to about 40 cents per Watt now (Figure 13.7). Four countries – Morocco, Chile, Egypt, and Mexico – are producing solar power at a cost of less than 3 cents per kilowatt hour, a price that is *cheaper* than natural gas. Coal is becoming obsolete, yet still poses an existential threat to the planet, especially in developing nations where it is often the easiest and cheapest solution. Yet more than half the new energy production capacity for electricity is renewable, and wind and solar capacity is doubling every four years.[22]

[22] https://exponentialroadmap.org/.

Summarizing the Known Unknowns

What we know is that there are a lot of known unknowns about where Terra's life support system is heading. Let's conclude by considering estimated changes in the 2009–2018 global carbon budget, as recently analyzed by The Global Carbon Budget Project (Figure 13.8). The big thick arrows show estimates of the anthropogenic carbon dioxide fluxes. Greenhouse gas emissions and land use changes (deforestation) increase CO_2 fluxes into the atmosphere (+35 and +6 gigatons per year, respectively). These are partially offset by increased absorption by the land and ocean (about 12 and 9 gigatons per year, respectively). The difference between these inputs and extractions is the increase in atmospheric CO_2, a big +20 gigatons per year.

This is all plenty scary. But we are still rolling around on top of the Dixie® Cup. Notice the nice thin arrows in Figure 13.8 – the ones marked "carbon cycling $GtCO_2$ per year." These are natural exchanges between the atmosphere and land surface. And they are really big – about 440 and 330 gigatons per year for the land and ocean, respectively. These big numbers mean there is a big natural exchange in CO_2, an exchange we are deeply involved with every time we breathe or eat a carrot. On land, soils, permafrost, and plants store carbon.

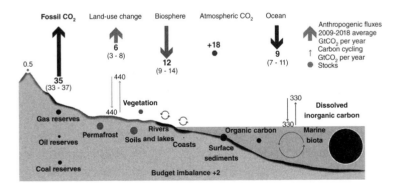

Figure 13.8 Human-induced changes in the 2010–2019 Global Carbon Budget, from the Global Carbon Budget.

In the ocean, sediments, plants, and animals store some carbon, but a whole lot of carbon also ends up dissolved in water. This dissolved carbon makes the ocean much more acidic.

There is a very real risk that global warming may muck up the basic balances that keep our Dixie® Cup planet centered on the up and up. On land, rising temperatures might kill soil microbes, or melt permafrost, or kill trees. Perhaps all three could happen together. These perils occur at scales and with complexities not captured well by our current models. So too we face very serious challenges when modeling glaciers and Arctic and Antarctic sea ice. These are highly nonlinear systems that we don't understand very well. And then there are temperature and ocean acidification impacts on the ocean, harming coral reefs, oysters, mussels, sea urchins, starfish, plants algae, and fish. Ocean acidification, for example, makes it harder for coral reefs to rebuild themselves, so that "coral reefs will transition to net dissolving before end of century."[23]

Each of these negative impacts are concerning taken alone. And each of these, and other unknown time bombs, might seriously shift the subtle balance of carbon dioxide and the Earth's energy balance in a potentially explosive manner. There could be catastrophic losses. Trees and reefs die. We know that during the very large 2015/16 El Niño, coral reefs faced unprecedented bleaching (i.e, death), while the Amazon rainforest suffered and the absorption of Co_2 diminished substantially. If such strong El Niños become more common or more intense, and occur in a world with a warmer ocean and atmosphere and a more acidic ocean, it is very plausible that we will both see rapid and lasting CO_2 increases. Pushed beyond a limit, ocean and tropical ecosystems fail. And they are likely to fail just when we need them most.

[23] http://science.sciencemag.org/content/359/6378/908.

14 WE CAN AFFORD TO WEAR A WHITE HAT

> It is better to light a candle than to curse the darkness.
> W. L. Watkinson[1]

Everybody likes Rick Blaine in *Casablanca*, Rooster Cogburn in *True Grit*, Han Solo in *Star Wars*: bad people who can't help but end up being good. Against their better judgment they do the right thing. We love their drama because it is our daily struggle. We are just regular girls and guys, not pretentious Jedi or suave Czechoslovakian resistance fighters. We want to keep our heads down, drive our cars, fly on planes, make and save money for our children.

But we also feel. And feel compassion. In an increasingly connected world, we are learning that climate change is making weather and climate extremes more intense and frequent, placing people in harm's way (Figure 14.1).

And we also have faith in our religion, our philosophy, our culture, our science. As the historian Yuval Harari has

---

[1] Watkinson, W. L., and H. Fleming. "The invincible strategy," in *The Supreme Conquest and Other Sermons Preached in America*. New York: Revell Company, 1907.

Figure 14.1 Lives in the balance. A malnourished child looks up at a scale as he is weighed and treated at Banadir hospital in Mogadishu. UN Photo/Tobin Jones, March 9, 2017, Mogadishu, Somalia, Photo # 716500

pointed out,[2] these "stories" that we tell ourselves permeate our lives, structuring our body politic. Stories inform how and if our civilization self-organizes and works, or doesn't, for good or for evil. Faith is the engine of history. We all believe in something. Some of us believe that there is little to believe in: best perhaps to just pursue wealth while promulgating doubt. Others assume that we are powerless to stop environmental destruction, best perhaps to just pursue wealth while promulgating liberal lamentations. Both of these perspectives are wrong. We don't need to believe in climate change, we can see and understand it. We don't need to accept climate change, we can radically transform our energy economies while improving our early warning and forecasting systems.

If you look back over human history, you will certainly find tragedy on catastrophic scales, but you also can't deny that modernity has worked out pretty well for many people. Yuval Harari tells us, "History is something that very few people have been doing while everyone else was plowing fields and carrying water buckets." Today most of us are free from lives of grueling labor, and we enjoy tremendous opportunities to express ourselves, learn, and recreate. For most of history many of our children died young, one of the most tragic of all losses. Just since 1991, the number of deaths for children under five has dropped by more than 50 percent[3] (Figure 14.2). In 1991, a stunning number of children in the world died before they were five: almost one out of every ten. By 2019, that number was still too high – one out of about twenty – but we have made substantial progress. Since 2000, this improvement has saved the lives of more than *50 million children*. That is about the same number of people who died fighting in World War II. If we act with compassion, guided by science, together we can achieve great things. We can share faith in our civilization, science, and humanitarian ideals. Hope leads us to work together to effect

---

[2] Harari, Yuval Noah. *Sapiens: A Brief History of Humankind*, New York: HarperCollins, 2015.
[3] https://data.unicef.org/topic/child-survival/under-five-mortality/.

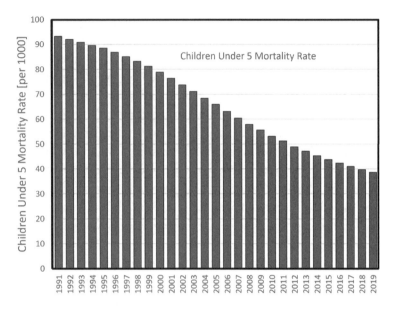

Figure 14.2 UNICEF child mortality data. The number of deaths per year per 1,000 children under 5. https://data.unicef.org/topic/child-survival/under-five-mortality/

positive change. Compassion for the poor, the hungry, and our beautiful fragile planet gives us courage – true grit sufficient enough to wear the White Hat. Rick Blaine, Rooster Cogburn, and Han Solo teach us that being blameless is not good enough. Maybe it's better to drink whiskey, fight, and yet be actively good. Action links faith and change, producing progress.

Famine, or the lack thereof, provides a sterling example of what we can achieve. At the close of the nineteenth century, famines wracked Europe and Africa. In India and China, more than 30 million people lost their lives.[4] In the 1940s, 1950s, and 1960s, more than 43 million people perished of famine.[5] Today, despite the existence of a massive number of extremely hungry people – more than 88 million in 2020 – famine-related

---

[4] Davis, Mike. Late Victorian Holocausts: El Niño Famines and the Making of the Third World. Verso Books, 2002.

[5] Hasell, Joe, and Max Roser. "Famines." Published online at OurWorldInData.org (2020). Retrieved from: https://ourworldindata.org/famines.

Figure 14.3 German carbon dioxide emissions. Data based on the Global Carbon Project archive.

mortality rates are very low. This reduction, achieved through effective international collaborations that connect the power of Earth science with humanitarian relief agencies is "one of the great unacknowledged triumphs of our lifetime."[6]

Just as humanity has tackled famine and infant mortality, so too fighting climate change is very much in reach. We can be both *wealthy* and *wise*. For example, Germany has invested in a highly efficient, high-tech, highly renewable economy. Figure 14.3 shows a time series of annual German carbon dioxide emissions. Since a peak in 1979, emissions have dropped by one-third to about 207 megatons of CO_2 in 2018. In that same time period, Germany's gross domestic product increased by more than 400 percent, going from US$878 billion to US$4 trillion in 2018. In 2019, renewable energy sources accounted for more than 34 percent of Germany's electricity production.

---

[6] de Waal, A. "The end of famine? Prospects for the elimination of mass starvation by political action." *Political Geography* (2017).

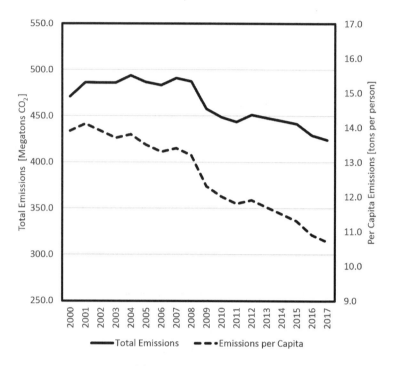

Figure 14.4 California Emissions and Per Capita Emissions.

Investments in education, infrastructure, and worker education created a sustainable, lucrative, and equitable economy in which many people have good paying jobs.

In the United States, the state of California provides another relevant example.[7] Despite a rapidly growing economy, Californian investments in alternative energy and commitments on minimizing emissions have resulted in substantial emission declines since 2000 (Figure 14.4). Per capita emissions declined by about 24 percent from peak levels, and actual emissions by about 14 percent.

We can also look at California's emissions as a function of gross domestic product (GDP) (Figure 14.5). "Energy Intensity" measures the amount of CO_2 emissions needed to

[7] https://ww3.arb.ca.gov/cc/inventory/pubs/reports/2000_2017/ghg_inventory_trends_00-17.pdf

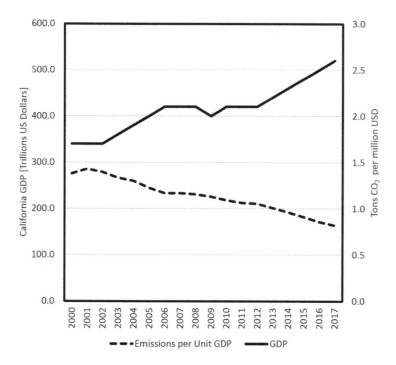

Figure 14.5 California GDP and energy intensities.

produce a certain amount of GDP. Since 2001, California's energy intensity has declined by 38 percent. Over the same time period, the economy experienced a massive growth of 41 percent. Investments in green technologies and increased efficiency can be completely compatible with rapid economic growth.

Why Our Human "Time Bomb" Can Be a Good Thing

We should build history together.

We should realize that we face an exciting "time bomb," rife with potential. The twenty-first-century "population explosion" carries within it the seeds of our own resurrection.

Each life is frantically *unique*. Like the Little Prince in Antoine De Saint-Exupéry's story, who despairs when he realizes that there are thousands of roses identical to the one he loves, we often forget how magically distinctive, utterly irreplaceable, and existentially exceptional each human life is. We all have our own unique past, our worldview, our own thoughts, emotions, and insights. Within our diverse and increasingly educated billions, we hold a tremendous ability to innovate and self-organize, to tell stories and create.

But being alive is only a necessary, but not sufficient, cause of consciousness. Having a life is not the same as *owning* your life. Living, by itself, does not make you conscious. For that, we need to pay careful attention to both the external world and the world within. Of course, that requires effort, but that effort brings rewards. With luck, each of us may have the gift of about three billion precious seconds – 90 years × 365 days × 24 hours × 60 minutes × 60 seconds. How many of these seconds will we claim as ours? This may make all the difference. Are we doing or being done to? Are we paying attention to the world around us, to Nature and our fellow travelers on Spaceship Earth? Or do we fritter away our moments consumed by consumption, competition, and distraction? As the climate crisis draws us all together, inattention may truly pose an existential threat.

As we have seen in the previous thirteen chapters, the rate of climate change is alarming. We are already experiencing extreme impacts. But our potential for effective intervention is also immense.

I like to think of this potential as a "time bomb," with "bomb" used in a positive sense of explosive potential. As

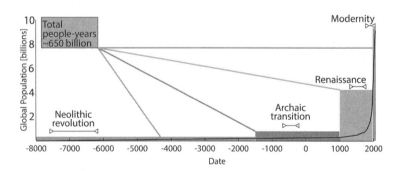

Figure 14.6 Global population and equivalent quantities of human people-years.

Figure 14.6 illustrates, between 1961 and 2050, humanity's time bomb will execute humanity's greatest experiment in parallel processing, as billions of individuals grow, think, discover, and consume. We will do far more, learn far more, and consume far more than any previous generation. Between 1961 and 2050, the Earth's population will triple, going from approximately 3 billion in 1961 to more than 9 billion in 2050.[8] As our population rapidly expands, however, our industry and technology will also enlarge rapidly. So we are currently in the middle of a very exciting, very fraught, explosion of human activity. Many things will be happening at once as we accelerate our consumption, creativity, and interconnectedness. Paying attention right now, seeing the role of climate change in the world today, is critical. We can let this mess run unfettered or, like Germany and California, we can grow wise but continue growing.

We can appreciate this urgency by breaking all of human history into four equivalent blocks of people-years, or "human capacity" (Figure 14.6). The y-axis represents global population estimates from 8000 BC through 2050 AD. The "hockey stick" structure clearly shows the rapid population growth of the modern era. That rapid growth is certainly placing strains on our planet. What is less appreciated, however, is how greatly we are expanding our human capacity (for good or ill). In analogy with the idea of "horsepower," we can define human capacity as the total number of people-years over a given span of time. Human capacity is calculated by adding up all the people living on the planet each year. Adding up all the people-years between 1960 and 2050, we get an estimated total of 631 billion person-years. This staggering amount of time is about forty-nine times the age of our universe (~13 billion years).

Another way to contemplate this temporal mass of humanity is to see it as roughly equivalent to the total number of global person-years between 1000 and 1960. The 90 years between 1961 and 2050 will see as much human capacity as

---

[8] Based on the United Nations medium variant projections.

approximately the 960 years prior (1000 AD to 1959). If we go back even further, in the two and a half millennia between 1500 BC and 1000 AD we find about 631 billion person-years. Finally, in the 6,500 years between 1500 BC and 8000 BC, approximately 631 billion person-years brings us back to time of the Neolithic revolution that saw the first continuous human settlements, the advent of farming, and the first domestication of animals.

Thus in just 90 years, 1961–2050, the Earth will witness an explosion of human activity broadly equivalent to the previous 960, 2,500, and 6,500 years before that. Furthermore, we are now mostly industrialized, far wealthier, far better educated, linked electronically, computer-enhanced, and sensor-enabled. We can also build on the knowledge of all the previous generations. A lot is likely to happen very quickly.

We will be able to wear the White Hat, if we want to pay attention. Over the past fifty years, our creative capacities have improved as education expanded, farmers became more productive, and research, development, and entrepreneurialism spread into Asia, South America, and Africa. There is good news on many fronts. Rates of extreme poverty are falling. The number of people without access to water has dropped by 50 percent since 1990. Overall global per capita production (a measure of our total economic production divided by the total number of people) has gone from $450 (in 2017 dollars) per person in 1960 to $11,312 in 2017. That is a stunning 21-fold increase in global per capita production – a measure of the power of our three billion seconds in action. The increase in total global gross domestic product has been incredible, and now stands at nearly $90 trillion a year (Figure 14.7), doubling just since 2004.

The world's rapid economic growth, however, is also a measure of our two-sided "time bomb." Our increased wealth (Figure 14.7), combined with our increased population (Figure 14.6), has led to massive increases in greenhouse gas emissions. Yet on the other side of the coin we could be ready to cut away inefficiency, develop new green technologies, and build

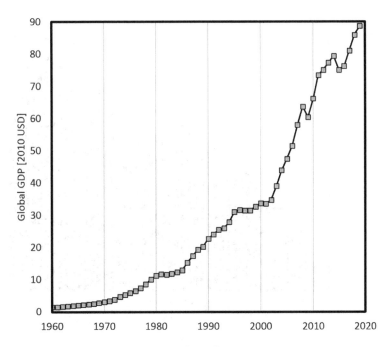

Figure 14.7 World Bank estimates of global gross domestic product in 2017 US dollars.

a more sustainable future, as Germany and California are attempting to do. And as our creativity and interconnections increase, magnified by rapid increases in global education and scientific research, and augmented by ever-faster computers, models, and sensors (satellites and various forms of microscopic scanners), we are already well under way to an "imaging revolution." Never have so many seen so much so well. We use satellites and sensors to probe the edges of time and the limits of space. We also use satellites and sensors to map and monitor our living world. We are also, as never before, teaching ourselves to learn.

Since 1970, we have seen the proportion of children receiving primary education rise from 72 percent to 89 percent (Figure 14.8). Not only has education expanded in general, but the gap between primary education for girls and boys has disappeared (at least globally, on average). Yet at more advanced

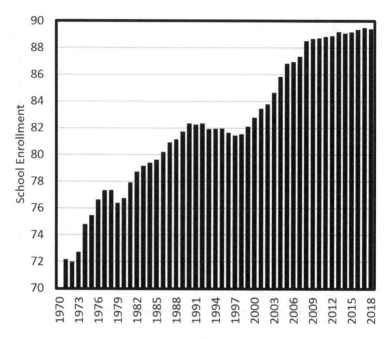

Figure 14.8 Percent of children receiving primary school education.
Source: World Bank

education levels, women in poorer countries still face substantial gaps, and women from the poorest families still face the highest barriers to education.

And as we have grown better educated, we have become more innovative. As wealth and education spread, the world has produced many more scientists, engineers, and inventors. Just since 1985, the number of global patent applications has quadrupled (Figure 14.9), reaching more than two million in 2016, a figure twice that of a decade before. Together, we are productive, educated, and innovative in ways our early ancestors could not even begin to imagine.

But now we have come to an era in which our very success may be our undoing. There is a *lot* going on, very *rapidly*. This incredible increase in productivity may come at a very steep price – because there has been a very strong coupling

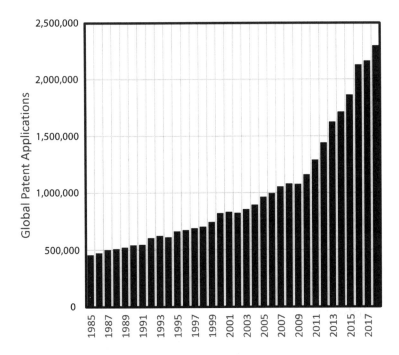

Figure 14.9 Global patent applications.
Source: World Bank

between our increase in production and the increase in our greenhouse gas emissions. But hope for the cure may be embedded in the cause. Lifted out of abject poverty in mere decades, our global billions certainly have the capacity to effectively tackle climate change. Given the will.

The need for this effort is profound. Right now we appear to be headed for 3°C of warming or more. This level of warming would almost certainly have catastrophic and potentially irreversible impacts on our planet's life support system. This book has documented the dangerous and expensive increase in extremes associated with just the 1°C warming we have already experienced to date. At +3°C or +4°C, the health impacts of heat waves would become terrible killers for billions of people in hot, humid regions. Crop production in the tropics, and especially Africa, would be dramatically impacted. A +3°C or +4°C world will bring much more frequent superstorms, riding across a rising tide of higher sea levels. And then there might be very real risks associated with positive feedbacks: melting ice caps, dead coral reefs, acid deoxygenated seas, and reduced carbon uptake from tropical forests could lead (very easily) to runaway warming.

Even the difference between 1.5°C and 2°C of warming will make an incredible difference. According to the Intergovernmental Panel on Climate Change's *Special Report on the Impacts of Global Warming of +1.5°C*,[9] having a total 2°C of warming versus 1.5°C would:

- Double the number of people exposed to extreme heat waves (limiting the total to 15% as opposed to 30%);
- Increase the area of melting permafrost by about two million km^2. This is an area four times the size of France;
- Double or triple the number of species experiencing severe habitat loss;
- Greatly increase the risk of irreversible loss of sea ice;
- Greatly increase the damage to coral reefs and fisheries via the impacts of very warm ocean temperatures, acidification, and deoxygenation;
- Increase the chance of an ice free Arctic ocean tenfold;
- Increase the number of people exposed to water stress by 50 percent;
- Increase by hundreds of millions the number of poor, food-insecure people exposed to climate risks.

The data we have explored together in this book strongly supports these concerns. As we saw in Chapter 4, the 2015–2019 time period has been a lot warmer than even the 1990s, about 0.7°C degrees warmer on land (on average). But this warming is not spread evenly, but rather acts in a more dangerous way. At any given time we find areas of exceptionally warm land and ocean temperatures (Figure 5.3); the fraction of the Earth with exceptional temperatures has tripled from the early 2000s,

---

[9] The International Panel on Climate Change Special Report: "Global Warming of 1.5°C," www.ipcc.ch/sr15/.

going from 6 percent to 18 percent. This warming, combined with population growth, has led to a stunning increase in the number of people exposed to heat waves each year (Figure 5.6).

In Chapters 8, 9, and 10 we have explored the dangerous implications of a 0.5°C to 0.7°C warming in the tropical oceans. This "extra" warming can exacerbate the intensity of naturally occurring climate phenomena like the Indian Ocean Dipole, El Niños, and La Niñas. Fore-armed with this conceptual model of climate change, however, we can translate these superwarm anomalies into opportunities for prediction. "Storyline" approaches to climate attribution have linked climate change to deadly droughts that have already helped push tens of millions more people into food insecurity, risking a reversal of decades of progress.[10] With our own eyes, we have seen how climate change made the 2015–2016 El Niño and the 2016–2017 La Niña more dangerous.

And with our own eyes we have seen how the most intense rainfall events are increasing in intensity (Figure 6.5), and how these extremes have affected hundreds of millions of people just since 2015 (Table 6.1). Focusing on the United States, where there is adequate data to track disasters quite precisely over time, we find huge increases in hurricane damages (Figure 7.4) and wildfire extent (Figure 11.3).

But seeing alone is not enough. We need to understand. To this end we have explored and explained how the Earth's radiation balance (and imbalance) works. Absorbing and re-emitting the sun's energy, the Earth provides a "fragile flame" that translates solar energy into complex patterns, miraculous life, and billions of astounding human stories. We are messing heavily with the life support system of our living planet, the only planet we know of that supports life. There is no planet B.

One recurring theme has been how incredibly thin the Earth's atmosphere is. While the sky seems vast when we look up, it is rather empty; there just isn't that much there up there.

---

[10] The 2018 United Nations Food and Agriculture Organizations annual "The State of Food Security and Nutrition Report." www.fao.org/3/I9553EN/i9553en.pdf

302 / Drought, Flood, Fire

Figure 14.10 The Earth as seen from the International Space Station.
Source: NASA astronaut Scott Kelly

For me, pictures taken by humans from the International Space Station evoke this understanding. Knowing that a human heart and mind reside behind the lens evokes a sense of intimacy. Taken from the edge of space, these photos illustrate the beauty of Earth but also the shallowness of our atmosphere. NASA astronaut Scott Kelly captured a beautiful example[11] on September 22, 2015 (Figure 14.10). On the left of this image the Nile River stretches across Egypt and Sudan. The band across the top shows the "airglow" at the very topmost layer of our atmosphere – the ionosphere.[12]

Seen from this perspective, we can begin to understand why dumping 37 gigatons of carbon dioxide a year into our atmosphere is like five people chain-smoking in a small car with the windows rolled up. That is not going to end well, and it is not going well now. Increases in human exposure to weather-related dangers are combining with increased climate volatility,

---

[11] www.nasa.gov/image-feature/the-nile-at-night.
[12] https://ourplnt.com/secrets-earth-multi-colored-airglow/#axzz6FTaOH9eP

costing us billions of dollars and yuan and euros and rupees each year, and impacting millions of people each year. So what might it cost us to seriously tackle climate change? The IPCC *Impacts of Global Warming of +1.5°C* report looked at this issue. While a detailed treatment is beyond the scope of this book, we can conclude *Drought, Flood, Fire* with a brief synopsis of the report's suggestions:

- Reducing carbon and methane emissions by 45 percent by 2030, reaching *zero emissions* by 2050;
- Switching very rapidly to renewable energy sources;
- Rapidly improving the efficiency with which we use energy;
- Stopping the burning of coal;
- Switching to biofuels;
- Planting billions of trees to absorb more CO_2;
- Increasing investments in low-carbon energy sources and increased energy efficiency;
- Taking an "all of the above" approach – simultaneously pursuing reductions in emissions, increases in efficiencies, and increases in terrestrial carbon uptake – is essential.

Can we afford to make these changes? Absolutely. It won't be cheap, but it is plausible, and our kids are worth it. According to the +1.5°C Special Report, "additional annual average energy-related investments for the period 2016 to 2050 ... are estimated to be around US$830 billion." This number is associated with changes that transition us to renewables while increasing our efficiency. This a big number, but is only about 1 percent of the total 2018 global gross domestic product (US$85 trillion). For reference, we can note that between 2001 and 2018, the United States is estimated to have spent $5.9 trillion on wars and military action in Afghanistan, Iraq, Syria, and Pakistan.[13,14] That works out to about $328

---

[13] https://watson.brown.edu/costsofwar/files/cow/imce/papers/2018/Crawford_Costs%20of%20War%20Estimates%20Through%20FY2019%20.pdf.
[14] www.cnbc.com/2018/11/14/us-has-spent-5point9-trillion-on-middle-east-asia-wars-since-2001-study.html.

billion per year, or about 1.5 percent of our estimated US 2018 GDP of US$20.5 trillion. The global costs of weather-related disasters in 2017 and 2018 have been estimated to be similar in magnitude – about $325 billion in damages each year.[15] The point here is not to advocate specific mitigation approaches, but rather to point out that the data tell us that we are already essentially "at war" with the climate. The magnitude of the economic and humanitarian impacts of this war are already large, and are almost certainly going to increase. But the cost of acting is relatively modest. What's $830 billion between friends in a $90-trillion global economy? We will end up paying one way or another, and the right thing to do, from a moral, economic, and environmental perspective, is to pay now for an inevitable energy transition.

Here is where the stories we tell each other matter, coordinating our intercontinental "time bomb" into explosive positive change. We need to act coherently and quickly to avoid rapid warming. We need to see and understand climate change as something that is hurting people and our planet *now*. We have been blessed with a miraculous planet, and the opportunity to live through an exciting and momentous time, filled with increasing prosperity and innovation. We *can* wear the White Hat. Compassion and logic *demand* that we do so. The great spiritual leaders taught us that "peace is a verb," telling us stories that eventually led to unprecedented peace and prosperity.[16] We need to make peace with our planet. Climate change is hurting people and our planet; we need to act to prevent that, and we *can* afford to do so.

---

[15] http://thoughtleadership.aonbenfield.com/Documents/20190122-ab-if-annual-weather-climate-report-2018.pdf.
[16] Pinker, Steven. The better angels of our nature: Why violence has declined. Penguin Group USA, 2012.

APPENDIX
A Few Resources for Further Reading and Research

Please note that this book is intended as an accessible starting point to help readers learn about recent climate extremes. While it is not my intent to review or rigorously represent the current scientific literature, below I provide brief sections that provide pointers for you to delve deeper into the various topics covered in this book.

Chapter 1

The Intergovernmental Panel on Climate Change (IPCC) Reports[1] provide extensive and authoritative climate change analyses compiled by panels of international experts. Recent reports on the dangerous impacts of +1.5°C and +2°C of warming, climate change and land,[2] and the ocean and cryosphere in a changing climate[3] provide valuable assessments of climate change threats. The forthcoming 6th IPCC Assessment Report[4] will contain an exhaustive and detailed analysis with contributions from hundreds of the best climate scientists. The

---

[1] Intergovernmental Panel on Climate Change (IPCC), www.ipcc.ch/
[2] IPCC Special Report on Climate Change and Land, www.ipcc.ch/report/srccl/
[3] IPCC Special Report on the Ocean and Cryosphere in a Changing Climate, www.ipcc.ch/report/srocc/
[4] IPCC 6th Assessment Report - www.ipcc.ch/report/sixth-assessment-report-cycle/

annual issues of the Bulletin of the American Meteorological Society on *Explaining Extreme Events from a Climate Perspective*[5] provide timely analyses intended for the general public. Each year's special issue provides a series of short articles that analyze the potential role that climate change played in a single extreme climate event. The National Academy of Sciences' *Attribution of Extreme Weather Events in the Context of Climate Change* provides an excellent overview of extreme event attribution. The International Detection and Attribution Group,[6] the World Climate Research Program's Grand Challenge on Weather and Climate Extremes,[7] the distributed ClimatePrediction.net,[8] the Climate Central World Weather Attribution Group,[9] the World Weather Attribution Group,[10] and many other excellent research teams constantly publish research on extreme attribution.

For those interested in a good introduction to climate change, Jeffrey Bennett's *A Global Warming Primer*[11] is a good place to start, as is Joseph Romm's *Climate Change: What Everyone Needs to Know*.[12] The edited volume *Climate Extremes: Patterns and Mechanisms*[13] has articles on many recent topics.

---

[5] Bulletin of the American Meteorological Society Annual Explaining Extreme Event Special Issues, www.ametsoc.org/ams/index.cfm/publications/bulletin-of-the-american-meteorological-society-bams/explaining-extreme-events-from-a-climate-perspective/

[6] Website of the International climate change Detection and Attribution Working Group, www.clivar.org/clivar-panels/etccdi/idag/international-detection-attribution-group-idag

[7] World Climate Research Program Extremes Events program, www.wcrp-climate.org/gc-extreme-events

[8] Website for the climateprediction.net project, www.climateprediction.net/

[9] Climate Central World Weather attribution page, www.climatecentral.org/go/wwa

[10] The World Weather Attribution Group website, www.worldweatherattribution.org/

[11] Bennett, Jeffrey O. *A Global Warming Primer: Answering Your Questions about the Science, the Consequences, and the Solutions.* Big Kid Science, 2016.

[12] Romm, Joseph. *Climate Change: What Everyone Needs to Know.* Oxford University Press, 2018.

[13] Wang, S-Y. Simon, et al., eds. *Climate Extremes: Patterns and Mechanisms.* Vol. 226. John Wiley & Sons, 2017.

Chapter 2

Chapter 2 draws heavily from Jan Zalasiewicz and Mark Williams's excellent book *The Goldilocks Planet: The 4 Billion Year Story of Earth's Climate*.[14] Fred Adam's *Origins of Existence: How Life Emerged in the Universe*[15] also informed this discussion. Several *Scientific American* articles by Caleb Scharf, a British-born astronomer and the director of the multidisciplinary Columbia Astrobiology Center at Columbia University in New York, helped guide the discussion of Blue, Green, and Red galaxies. You might be interested in How *Black Holes Shape the Galaxies, Stars and Planets around Them*[16] or *Is Earth's Life Unique in the Universe?*[17] NASA provides a great summary of Cosmology and the Big Bang Theory at map.gsfc.nasa.gov/universe/. The El Niño-Southern Oscillation blog posts (www.climate.gov/news-features/blogs/enso/index-page-enso-blog-posts), written by Michelle L'Heureux, Nat Johnson, and Tom Dilberto, featured (briefly) in Figure 2.6, provide both good background material and interesting up-to-date analyses. *Global Physical Climatology*[18] by Dennis Hartmann provides a great overview of the General Circulation. *Meteorology Today: An Introduction to Weather, Climate, and the Environment, 9th Edition*[19] by Donald Ahrens is a deservedly popular treatment of these topics in accessible format.

[14] Zalasiewicz, Jan, and Mark Williams. *The Goldilocks Planet: The 4 Billion Year Story of Earth's climate*. Oxford University Press, 2012.
[15] Adams, Fred C. *Origins of Existence: How Life Emerged in the Universe*. Simon and Schuster, 2010.
[16] Scharf, Caleb, Black Holes Shape the Galaxies, Stars and Planets around Them, www.scientificamerican.com/article/how-black-holes-shape-galaxies-stars-planets-around-them/.
[17] Scharf, Caleb, Is Earth's Life Unique in the Universe?, www.scientificamerican.com/article/is-earth-s-life-unique-in-the-universe/.
[18] Hartmann, Dennis L. *Global Physical Climatology*. Vol. 103. Newnes, 2015.
[19] Ahrens, C. Donald. Meteorology *Today: An Introduction to Weather, Climate, and the Environment*. Cengage Learning, 2012.

Chapter 3

This chapter (like Chapter 2) draws heavily from Jan Zalasiewicz and Mark Williams's excellent book *The Goldilocks Planet: The 4 Billion Year Story of Earth's Climate* and Fred Adam's *Origins of Existence: How Life Emerged in the Universe*. *Global Physical Climatology* by Dennis Hartmann and *Meteorology Today: An Introduction to Weather, Climate, and the Environment, 9th Edition*, by Donald Ahrens provide great introductions to climatology and meteorology. The book *Atmospheric Thermodynamics* by Craig F. Bohren and Bruce A. Albrecht,[20] while very technical and not for the faint of heart, is excellent, and has inspired certain aspects of this work, including my attempts to mix humor and science.

Chapter 4

Following up on the DIY aspect of Chapter 4, I will mention here some of the many excellent online resources for examining long time series of weather and climate data. Please note, however, that this is not intended to be an authoritative review. I am simply listing some of the resources that I have used in this book. The KNML climate explorer (climexp.knmi.nl) provides access to a vast climate archive. This access portal hosts a large number of observational data sets, as well as a large number of climate change simulations, making it an exceptional resource for the community. NOAA's Earth System Research Laboratory has a rich set of online tools (www.esrl.noaa.gov/psd/cgi-bin/data/getpage.pl), and the International Research Institute's Data Library (iridl.ldeo.columbia.edu/index.html) is a powerful and freely accessible online data repository and analysis tool that allows a user to view, analyze, and download hundreds of terabytes of climate-related data through a standard web browser.

---

[20] Bohren, Craig F., and Bruce A. Albrecht. *Atmospheric Thermodynamics*. Oxford University Press, 2000.

Chapter 5

The following resources are likely to be useful to interested readers. The World Climate Research Program group on Weather and Climate Extremes website in general, and their special issue in the journal *Weather and Climate Extremes* in particular,[21] provides a valuable collection of articles. A 2019 Lancet article[22] provides a good overview of current impacts of temperature extremes and the literature related to them. In general, this chapter has broadly followed the approach used by the Lancet article to estimate heat exposure events, though there are many different ways heat exposure indices can be estimated. It should also be noted that this chapter did not attempt to discuss in detail many other important aspects of temperature extremes, such as negative crop impacts, wildfire risk, reductions in worker productivity, increases in conflict, and increased disease transmission rates. The 2019 Intergovernmental Panel on Climate Change reports on *Climate Change and Land* and *Global Warming of 1.5°C*[23] provide relevant material discussing these impacts, as will the *6th Assessment Report on Impacts, Adaptation and Vulnerability*.[24]

Chapter 6

Once again, The World Climate Research Program group on Weather and Climate Extremes website in general, and their special issue in the journal *Weather and Climate Extremes* in particular, will be of value to interested readers. In this chapter, we followed the general approach used in a

---

[21] Weather and Climate Extremes, www.sciencedirect.com/journal/weather-and-climate-extremes/.

[22] Watts, Nick, et al. "The 2019 report of The Lancet Countdown on health and climate change: ensuring that the health of a child born today is not defined by a changing climate." *The Lancet* 394.10211 (2019): 1836–1878.www.thelancet.com/journals/lancet/article/PIIS0140-6736(19)32596-6/fulltext.

[23] IPCC, *Special Report: Global Warming of 1.5°C*, www.ipcc.ch/sr15/.

[24] IPCC. *6th Assessment Report on Impacts, Adaptation and Vulnerability*, www.ipcc.ch/report/sixth-assessment-report-working-group-ii/.

seminal *Nature Climate Change* study by Markus Donat and coauthors.[25] A 2019 special issue of *Environmental Research Letters* provides a useful collection of recent research,[26] and a 2013 edited volume *Extremes in a Changing Climate* provides an excellent set of technical articles on extremes. The annual *Global Assessment Reports on Disaster Risk Reduction* (GAR)[27] are the flagship reports of the United Nations on worldwide efforts to reduce disaster risk. The GAR is published biennially by the UN Office for Disaster Risk Reduction (UNDRR), and is the product of the contributions of sovereign, public, and private disaster risk-related science and research, among others. The 2019 GAR report discusses weather extremes, climate change, and disaster risk management. Disaster statistics can be obtained from the EM-DAT database, the US Billion Dollars disasters database,[28] and the Munich Re natural catastrophe database.[29] Brief news reports on flood events are available at floodlist.com.

Chapter 7

Two excellent attribution studies by Wehner and Risser[30] and Patricola and Wehner[31] were featured in this chapter.

---

[25] Donat, Markus G., et al. "More extreme precipitation in the world's dry and wet regions." *Nature Climate Change* 6.5 (2016): 508.

[26] *Environmental Research Letters* Focus on Extreme Precipitation, iopscience.iop.org/journal/1748-9326/page/Focus_on_Extreme_Precipitation_Observations_and_Process_Understanding.

[27] United Nations Office for Disaster Risk Reduction, 2019 Global Assessment Report on Disaster Risk Reduction, www.gar.undrr.org/sites/default/files/reports/2019-05/full_gar_report.pdf.

[28] NOAA Billion Dollar Disasters, www.ncdc.noaa.gov/billions/

[29] Munich Re Database of Catastrophic Disasters, natcatservice.munichre.com/.

[30] Risser, Mark D., and Michael F. Wehner. "Attributable human-induced changes in the likelihood and magnitude of the observed extreme precipitation during Hurricane Harvey." *Geophysical Research Letters* (2017).

[31] Patricola, Christina M., and Michael F. Wehner. "Anthropogenic influences on major tropical cyclone events." *Nature* 563.7731 (2018): 339. www.nature.com/articles/s41586-018-0673-2.

The two multi-authored synthesis reports prepared by the World Meteorological Organization Task Team on Tropical Cyclones and Climate Change provide a useful synopsis of current research. These reports focus on observations[32] and model-projected changes in tropical cyclone activity for a 2°C anthropogenic warming[33]. The National Academies report on extreme event attribution[34] describes the complexities surrounding the analysis of landfalling hurricanes and cyclones, and provides a good introduction to general topic of climate extreme attribution. Kevin Trenberth et al.'s 2015 article introduces the idea of "storyline" attribution,[35] and Lloyd and Oreskes[36] provide a very thoughtful and accessible discussion contrasting this approach to more traditional odds-based attribution approaches, which builds on Shepard's 2016 treatment of this subject.[37] How we frame our questions matters.

Chapter 8

Readers interested in finding out more about the emerging "storyline" approach to climate attribution may be interested in the accessible and interesting paper *Climate change attribution: When is it appropriate to accept new methods?* By

[32] Knutson, Thomas, et al. "Tropical Cyclones and Climate Change Assessment: Part I. Detection and Attribution." *Bulletin of the American Meteorological Society* 2019 (2019). journals.ametsoc.org/doi/pdf/10.1175/BAMS-D-18-0189.1.

[33] Knutson, Thomas, et al. "Tropical cyclones and climate change assessment: Part II. projected response to anthropogenic warming." *Bulletin of the American Meteorological Society* 2019 (2019). journals.ametsoc.org/doi/pdf/10.1175/BAMS-D-18-0194.1.

[34] National Academies of Sciences, Engineering, and Medicine. *Attribution of Extreme Weather Events in the Context of Climate Change.* National Academies Press, 2016.

[35] Trenberth, K. E., J. T. Fasullo, and T. G. Shepherd. "Attribution of climate extreme events." *Nature Climate Change,* 5 (2015), 725–730. doi.org/10.1038/nclimate2657.

[36] Lloyd, Elisabeth A., and Naomi Oreskes. "Climate change attribution: When is it appropriate to accept new methods?." *Earth's Future* 6.3 (2018): 311–325. doi.org/10.1002/2017EF000665.

[37] Shepherd, Theodore G. "A common framework for approaches to extreme event attribution." *Current Climate Change Reports* 2.1 (2016): 28–38. link.springer.com/article/10.1007%2Fs40641-016-0033-y.

Elisabeth Lloyd and Naomi Orestes. This paper contrasts more traditional "odds-based" attribution approaches with newer "storyline" approaches, developed by scientists such as Trenberth, Fasullo, and Sheperd. An accessible 2019 overview of this topic, quoting Stephanie Herring and Elisabeth Lloyd, is available at www.slate.com.[38] Theodore Sheperd[39] has also a led a relevant 2018 study on this topic published in *Climatic Change*. In Chapter 7 we have already seen a very relevant application and contrast of odd-based and storyline approaches applied to hurricanes and tropical cyclones. In Chapter 8, consistent with the energy-centric focus of this book, I have linked the increasing energy content of the oceans to potentially dangerous natural climate fluctuations like the Indian Ocean Dipole. NOAA scientists LuAnn Dahlman and Rebecca Lindsey have written an excellent blog describing the link between climate change and ocean heat content.[40] Figure 8.1 is an update of the results presented in their report. The Australian Bureau of Meteorology provides a good overview of the Indian Ocean Dipole,[41] as well as a video summary[42]. My concluding thoughts on Galileo draw from *The Case of Galileo: A Closed Question?* by Annibale Fantoli, translated by George V. Coyne.[43]

Chapter 9

These notes are intended to guide nonspecialists interested in learning more about El Niño, El Niño impacts, and

---

[38] www.slate.com/technology/2019/12/attribution-science-field-explosion-2010s-climate-change.html.

[39] Shepherd, Theodore G., et al. "Storylines: An alternative approach to representing uncertainty in physical aspects of climate change." *Climatic Change* 151.3–4 (2018): 555–571.link.springer.com/article/10.1007/s10584-018-2317-9.

[40] Dahlman, LuAnn, and Rebecca Lindsey. Climate Change: Ocean Heat Content, www.climate.gov/news-features/understanding-climate/climate-change-ocean-heat-content.

[41] Australian Bureau of Meteorology. Introduction to the Indian Ocean Dipole. www.bom.gov.au/climate/iod/.

[42] www.youtube.com/watch?v=J6hOVatamYs.

[43] Fantoli, Annibale. *The Case of Galileo: A Closed Question?* University of Notre Dame Press, 2012.

climate change. Mickey Glantz's *Currents of Change: Impacts of El Niño and La Niña on Climate and Society*[44] provides a wonderful introduction. Mike Davis's *Late Victorian Holocausts: El Niño Famines and the Making of the Third World*[45] provides a magisterial melding of global political history and global environmental history. More about my climate hero Sir Gilbert Walker can be found in Richard W. Katz's accessible historical analysis *Sir Gilbert Walker and a Connection between El Niño and Statistics*[46]. Motivated by the terrible late nineteenth-century El Niño–related famines that devastated India, Walker set out on an encyclopedic exploration of global weather that uncovered the global teleconnections that set the stage for much of modern climate science. Readers interested in finding out more about the recent impacts of the 2015–16 El Niño might be interested in the NOAA *Bulletin of the American Meteorological Society's (BAMS) Explaining Extreme Events* issues focused on 2015 and 2016. For those interested in marine ecosystems, the 2016 *BAMS* attribution issue has a number of articles focused on impacts of marine heat waves and their impacts on tropical marine ecosystems. Many of these results are summarized by Robert S. Webb and Francisco E. Werner in their article "Explaining Extreme Ocean Conditions Impacting Living Marine Resources. The annual United Nations State of Food Security and Nutrition in the World" (www.fao.org/publications/sofi/) documents the post-2015 increase in food insecurity and the need to better manage the effects of El Niño–related climate variability. The forthcoming International Panel on Climate Change *6th Assessment Report* will provide a valuable resource (http://www.ipcc.ch/assessment-report/ar6/). My personal research has explored the idea that a substantial proportion of the global warming signal

---

[44] Glantz, Michael H., *Currents of Change: Impacts of El Niño and La Niña on Climate and Society*. Cambridge University Press, 2001.
[45] Davis, Mike. *Late Victorian Holocausts: El Niño Famines and the Making of the Third World*. Verso Books, 2002.
[46] Katz, Richard W. "Sir Gilbert Walker and a connection between El Nino and statistics." *Statistical Science* (2002): 97–112.

can be broken into two components: a high-frequency component linked to stronger, warmer El Niños in the equatorial East Pacific; and a low-frequency "warming mode" linked to warming trends in the western Pacific and Indian Oceans.[47] A paper on this topic is available in the *Journal of Climate* (2015). A 2018 paper in the *Quarterly Journal of the Royal Meteorological Society* focuses explicitly on the 2015–2016 southern Africa drought (linked to El Niño) and the subsequent 2016–2017 East African droughts, which were tied to exceptionally warm West Pacific ocean conditions and La Niña.[48] In this chapter I also draw from two *Bulletin of the American Meteorological Society* attribution papers focused on climate change, the 2015–16 El Niño and extreme droughts and food insecurity in Ethiopia[49] and Southern Africa.[50]

Chapter 10

Please note: having spent twenty years studying East African rainfall has led to a longer treatment for this chapter.

There have been too many articles on the East African rainfall paradox and decline to list them all, but here is a historical synopsis with some of the highlights. The rainfall decline was first detected during routine efforts to monitor crop conditions in Ethiopia in 2003, and documented in a FEWS NET report in

---

[47] Funk, Chris C., and Andrew Hoell. "The leading mode of observed and CMIP5 ENSO-residual sea surface temperatures and associated changes in Indo-Pacific climate." *Journal of Climate* 28.11 (2015): 4309–4329.

[48] Funk C., L. Harrison, S. Shukla, C. Pomposi, G. Galu, et al. "Examining the role of unusually warm Indo-Pacific sea surface temperatures in recent African droughts." *Quarterly Journal of the Royal Meteorological Society* (2018). doi.org/10.1002/qj.3266.

[49] Funk, C., L. Harrison, S. Shukla, A. Hoell, D. Korecha et al. "Assessing the contributions of local and east Pacific warming to the 2015 droughts in Ethiopia and Southern Africa," *Bulletin of the American Meteorological Society* (December 2016): S75–S77, doi:10.1175/BAMS-16-0167.1.

[50] Funk C., F. Davenport, L. Harrison, T. Magadzire, G. Galu, et al. "Anthropogenic enhancement of moderate-to-strong El Niños likely contributed to drought and poor harvests in Southern Africa during 2016," *Bulletin of the American Meteorological Society*, 37 (2017): S1–S3, DOI. 10.1175/BAMS-D-17-0112.2.

2005 (pdf.usaid.gov/pdf_docs/PNADH997.pdf). I led a paper in 2008[51] that argued that this decline was tied to warming in the Indian Ocean. In 2010 and 2011, Park Williams and I published a USGS report and then a *Climate Dynamics*[52] paper in 2011 that extended the warming to the broader Indo-Pacific region. These studies documented the emergent dangers associated with recent La Niñas, and helped inform a successful FEWS NET drought forecast during 2010.[53] In 2012, Brad Lyon and Dave Dewitt published an important model-based study[54] that argued that Pacific sea surface temperatures played a dominant role in producing East African droughts. In 2013, Jessica Tierney and coauthors used paleo-climate data to link drying in East Africa to warming in the eastern Indian Ocean.[55] A similar analysis by Tierney and coauthors in 2015[56] also pointed out a potential partial solution to the East Africa Climate Paradox – the climate models had a tendency to dramatically underestimate the March–May East African rainy season. Several papers led by Andrew Hoell and Brant Liebmann in 2013[57] and 2014[58] emphasized the important role played by the interaction of the western and eastern Pacific. The year 2014 also saw important contributions by Wengchang Yang[59] emphasizing the important role of decadal variability in the

---

[51] Funk, Chris, et al. "Warming of the Indian Ocean threatens eastern and southern African food security but could be mitigated by agricultural development." *Proceedings of the National Academy of Sciences* 105.32 (2008): 11081–11086.

[52] Williams, A. Park, and Chris Funk. "A westward extension of the warm pool leads to a westward extension of the Walker circulation, drying eastern Africa." *Climate Dynamics* 37.11–12 (2011): 2417–2435.

[53] Funk, Chris. "We thought trouble was coming." *Nature* 476.7358 (2011): 7.

[54] Lyon, Bradfield, and David G. DeWitt. "A recent and abrupt decline in the East African long rains." *Geophysical Research Letters* 39.2 (2012).

[55] Tierney, Jessica E., et al. "Multidecadal variability in East African hydroclimate controlled by the Indian Ocean." *Nature* 493.7432 (2013): 389–392.

[56] Tierney, Jessica E., Caroline C. Ummenhofer, and Peter B. deMenocal. "Past and future rainfall in the Horn of Africa." *Science Advances* 1.9 (2015): e1500682.

[57] Hoell, Andrew, and Chris Funk. "The ENSO-related west Pacific sea surface temperature gradient." *Journal of Climate* 26.23 (2013): 9545–9562.

[58] Hoell, Andrew, and Chris Funk. "Indo-Pacific sea surface temperature influences on failed consecutive rainy seasons over eastern Africa." *Climate Dynamics* 43.5–6 (2014): 1645–1660.

[59] Yang, Wenchang, et al. "The East African long rains in observations and models." *Journal of Climate* 27.19 (2014): 7185–7202.

Pacific Ocean. Two 2014 papers led by Shraddhanand Shukla[60] demonstrated the predictability of the March–May rainy season, helping set the stage for the successful 2016 Climate Hazard Center forecasts. Two 2014 papers by Sharon Nicholson[61] documented the dangerous increase in recent droughts and explored the predictability of the March–May rainy season.

In 2015, David Rowell and coauthors published *Reconciling Past and Future Rainfall Trends over East Africa*,[62] the paper that termed the paradox phrase, concluding that natural variability "is unlikely to have been the dominant driver of recent droughts." In 2015, Andrew Hoell and I explored the idea that a substantial proportion of the global warming signal can be broken into two components: a high-frequency component linked to stronger, warmer El Niños in the equatorial East Pacific, and a low-frequency "warming mode" linked to warming trends in the western Pacific and Indian Oceans. In 2018, I led two efforts that (1) examined both the 2015–2016 southern African droughts and the 2016–2017 east African droughts; and (2) formally attributed the influence of climate change on the 2017 East African drought.[63] Finally, in their 2019 paper, "'Eastern African Paradox' rainfall decline due to shorter not less intense Long Rains,"[64] Caroline

---

[60] Shukla, Shraddhanand, et al. "A seasonal agricultural drought forecast system for food-insecure regions of East Africa." *Hydrology and Earth System Sciences* 18.10 (2014): 3907–3921; Shukla, Shraddhanand, Christopher Funk, and Andrew Hoell. "Using constructed analogs to improve the skill of National Multi-Model Ensemble March–April–May precipitation forecasts in equatorial East Africa." *Environmental Research Letters* 9.9 (2014): 094009.

[61] Nicholson, Sharon E. "A detailed look at the recent drought situation in the Greater Horn of Africa." *Journal of Arid Environments* 103 (2014): 71–79; Nicholson, Sharon E. "The predictability of rainfall over the Greater Horn of Africa. Part I: Prediction of seasonal rainfall." *Journal of Hydrometeorology* 15.3 (2014): 1011–1027.

[62] Rowell, David P., et al. "Reconciling past and future rainfall trends over East Africa." *Journal of Climate* 28.24 (2015): 9768–9788.

[63] Funk, Chris, et al. "Examining the potential contributions of extreme "western V" sea surface temperatures to the 2017 March–June East African drought." *Bulletin of the American Meteorological Society* 100.1 (2019): S55–S60.

[64] Wainwright, Caroline M., et al. "'Eastern African Paradox' rainfall decline due to shorter not less intense Long Rains." *NPJ Climate and Atmospheric Science* 2.1 (2019): 1–9.

M. Wainwright and coauthors have suggested that the recent decline is strongly associated with a shorter rainy season. During March, warmer waters to the south of East Africa delay the onset of the rainy season. During May, a decrease in surface pressure over Arabia supports an earlier cessation of the rainy season.

As this brief summary illustrates, substantial uncertainty continues, and the paradox remains only partially resolved. But we have come a long way.

Chapter 11

Readers interested in fire in the United States and elsewhere might be interested in Michael Kodas's compelling book *Megafire: The Race to Extinguish a Deadly Epidemic of Flame*. This book combines compelling narrative and science to describe both recent conflagrations as well as the complex issues surrounding these terrifying environmental disasters. In the United States, the Union of Concerned Scientists has assembled a good synopsis,[65] which was updated in March 2020. This page also links to a fascinating podcast by a fire expert, Professor John Bailey.[66] The Fourth US Climate Assessment[67] covers fires in its chapter on forests, and in the individual chapters on US climate regions.

Chapter 12

Boer, Resco de Dios, and Bradstock's accessible *Unprecedented Burn Area of Australian Mega Forest Fires* commentary[68] provided the main input for Chapter 12, augmented

---

[65] The Connection Between Climate Change and Wildfires, Union of Concerned Scientists, www.ucsusa.org/resources/climate-change-and-wildfires.
[66] Bailey, John. *The Science of Forest Fires: Culture, Climate, and Combustion*, www.ucsusa.org/resources/science-forest-fires-culture-climate-and-combustion.
[67] The Fourth United States Climate Assessment, nca2018.globalchange.gov/.
[68] Boer, Matthias M., Víctor Resco de Dios, and Ross A. Bradstock. "Unprecedented burn area of Australian mega forest fires." *Nature Climate Change* (2020): 1–2.

by two papers by Rachel Nolan and coauthors that describe and document the link between vapor pressure deficits, dead fuel moisture levels, and fire extent. Dr. Nolan's paper in *Remote Sensing of Environment*[69] describes the link between vapor pressure deficits and dead fuel moisture. Dr. Nolan's paper in *Geophysical Research Letters* describes the link between dead fuel moisture and wildfire extent.

This chapter has also drawn heavily on the *Summer of Crisis* report, by Lesley Hughes, Will Steffen, Greg Mullins, Annika Dean, Ella Weisbrot, and Martin Rice, which aptly describes the 2019–2020 fire season for New South Wales. Both this chapter and the *Summer of Crisis* report refer to the excellent reports produced by Australia's Bureau of Meteorology. Their site www.bom.gov.au/climate/ provides access to these reports, as well as more information about Australia's climate, the Indian Ocean Dipole (www.bom.gov.au/climate/iod/) and El Niño-La Niña impacts (www.bom.gov.au/climate/enso/).

Readers will also likely be interested in a detailed climate attribution study by Geert von Oldenburg and coauthors.[70] A summary of this excellent study can be found at the *World Weather Attribution* website www.worldweatherattribution.org/bushfires-in-australia-2019-2020/. The study focused on changes in fire weather over the area of the most intense bushfires, and concluded that climate did, and will, increase the probability of extreme fire weather. Readers are encouraged to access the full study at www.nat-hazards-earth-syst-sci-discuss.net/nhess-2020-69/.

---

[69] Nolan, R. H., V. Resco de Dios, M. M. Boer, G. Caccamo, M. L. Goulden, and R. A. Bradstock. "Predicting dead fine fuel moisture at regional scales using vapor pressure deficit from MODIS and gridded weather data." *Remote Sensory Environment* 174 (2016): 100–108.

[70] van Oldenborgh, G. J., F. Krikken, S. Lewis, N. J. Leach, F. Lehner, et al. "Attribution of the Australian bushfire risk to anthropogenic climate change," *Natural Hazards and Earth Systems: Scientific Discussions* (2020, in review), doi.org/10.5194/nhess-2020-69.

Chapter 13

Each year the Global Carbon Budget project puts out a fabulous assessment of the previous year's carbon emissions (www.globalcarbonproject.org/carbonbudget/). Observations of annual atmospheric CO_2 levels provided by Scripps (scrippsco2.ucsd.edu/) and NOAA (www.esrl.noaa.gov/gmd/ccgg/trends/global.html) provide a key input. The annual carbon budget is summarized in a yearly article in Earth System Science Data. The 2019 paper[71] is available at www.earth-syst-sci-data.net/11/1783/2019/. This paper describes the estimation process and major results. Another 2019 Global Carbon Project report, "Global Energy Growth is Outpacing Decarbonization,"[72] provides a concerning summary.

Chapter 14

This chapter draws on Yuval Noah Harari's excellent book *Sapiens: A Brief History of Humankind*.[73] A guiding principle of *Sapiens* is that shared stories have acted as a central innovation in human history, fostering coordinated activities that greatly leveraged our individual capabilities. Whether teaming up to take down a mastodon or build a corporation, stories provide a common framework that supports coherent action. This book has sought to reinforce such a framework by providing an accessible description of recent climate extremes, alongside accounts of associated humanitarian and economic impacts. As we face the dangers of climate change, we are also empowered to watch our evolving planet and humanity in ways that were never possible before. To this end, this chapter draws on the incredible data resources provided by the World Bank

---

[71] Friedlingstein, Pierre, et al. "Global carbon budget 2019." *Earth System Science Data* 11.4 (2019): 1783–1838.

[72] www.globalcarbonproject.org/global/pdf/GCP_2019_Global%20energy%20growth%20outpace%20decarbonization_UN%20Climate%20Summit_HR.pdf.

[73] Harari, Yuval Noah. *Sapiens: A Brief History of Humankind*. Random House, 2014.

(data.worldbank.org/indicator/). Another very valuable resource used here is provided by the annual United Nations Food and Agriculture Organization's *The State of Food Security and Nutrition* reports.[74] Since 2015, this data source indicates an increase in global hunger, related in part to increasingly extreme climate. Finally, the chapter also builds on natural catastrophe data summarized by AON Benfield.[75] AON Benfield, Munich Re,[76] and other re-insurance companies specialize in helping insurance companies manage risk. They have a profit-driven motivation to pay very close attention to climate extremes.

[74] www.fao.org/state-of-food-security-nutrition.
[75] thoughtleadership.aonbenfield.com/Pages/Home.aspx?ReportYear=2020.
[76] Munich Re Natural Catastrophe Database, natcatservice.munichre.com/.

INDEX

acute malnutrition, 189–190, 197
adenosine triphosphate, 39
anthropic principle, 21
arrow of time, 44
Arthur Eddington, 44
attribute, 112
attribution, 149
attribution analysis, 211, 261–262, 264–265
attribution science, 107
attribution study, 117–118, 232, 263, 318
Australian Bureau of Meteorology, 254–255, 312
available energy, 44, 47, 53

Bangladesh, 5, 9, 135, 151, 273
bathtub warming, 168
Berkeley Earth, 101
Big Bang, 23
black hole, 26–27
Black Summer, 267
Bulletin of the American Meteorological Society, 116, 127, 148, 171, 198, 204, 263, 273, 306, 311, 313–314, 316

Camp Fire, 11, 240, 243, 245, 248
Carnot Cycle, 43
Carr Fire, 243
Centre for Research on the Epidemiology of Disasters, 99, 198
China, 10, 99, 135–136, 152, 160, 190, 225, 280–281, 290
Clausius-Clapeyron, 127–128, 130
climate attribution, 13, 117, 153, 165, 204, 261, 263, 265, 301, 311, 318
climate attribution, 1, 12
climate change attribution, 149, 165, 311
climate change projections, 136
Climate Explorer, 63, 84, 137
climate hazard, 165
Climate Hazards Center, 101, 123, 131, 196, 214, 216
coal, 5, 35, 38, 279–280, 283, 303
Community Alert, 236
complexity, 16, 22, 39, 46–49, 148, 189
Convectively Available Potential Energy, 53, 55

coral bleaching, 9, 211, 273
coral reefs, 17, 80, 94, 114, 171,
    272–273, 280, 283, 286, 300
Coriolis, 34
counterfactual, 112, 118–119, 265
crop production, 4, 299
cumulonimbus, 56
cyanobacteria, 58

DanChurchAid, 1
dark energy, 23
dead fuel moisture, 256–258,
    263–266, 318
dead fuel moisture content, 258
detection, 149

Earth–Sun system, 16, 110
East Africa, 2, 9–10, 12, 18, 164,
    171, 173, 176, 179–180, 213,
    215, 222, 228–229, 232–233,
    315–316
East African Climate Paradox, 18,
    212–213
Egypt, 103, 191, 284, 302
Einstein, 69, 73
El Niño, 7, 9, 13–14, 18, 32, 77,
    186–187, 190, 192–195,
    197–198, 201–202, 204,
    206–210, 212–213, 216, 222,
    274, 286, 290, 301, 307, 312,
    318
El Niño–Southern Oscillation, 32,
    193
Emergency Events Database, 99, 136
energetically closed, 47
energetically open, 47
energy balance model, 71
Energy Convergence Model, 170
energy intensity, 292
entropy, 43–49, 51, 57
epidemiology, 117, 153
Ethiopia, 15, 18, 103, 164, 174,
    177, 189, 191, 195–196, 202,
    204, 210, 214, 219, 226, 232,
    314
event attribution, 12–13, 108, 153,
    306, 311
exposure, 7–8, 17, 37, 92–93, 101,
    103, 105–106, 109, 113–114,
    116, 265, 302, 309
external variations, 167

famine, 4, 10, 19, 86, 171, 189–191,
    196, 210, 213, 221, 290–291
famines, 14, 32, 190, 193, 290, 313
fingerprints, 5
fire extent, 257
flooding, 10, 132, 164, 171
food prices, 196, 198, 213, 221
Food Security and Nutrition
    Working Group, 4, 123
France, 99–100, 300

galaxies, 26
Galileo, 26, 37, 49, 183, 312
general circulation, 33
Global Carbon Budget, 123, 277,
    279, 282, 285, 319
Global Carbon Project, 68, 278,
    281–282, 291, 319
Goldilocks, 36
Goldilocks Planet, 16, 36, 74, 276,
    307–308
gravity, 23, 26
Great Barrier Reef, 9
greenhouse gas effect, 29, 65
gross domestic product, 152, 268,
    291, 296
Guinea, 103, 133

Hadley Circulation, 31, 34, 47, 53–54
heat death, 45, 47, 50–51
heat engine, 33, 45
heat wave, 99–100, 103, 105–106,
    113, 115, 117, 198
height fields, 222–223, 225

Heraclitus, 39, 67
holy moly, 175
human capacity, 295
Hurricane Harvey, 9, 17, 131, 145, 150, 154, 310
Hurricane Maria, 11, 18

increased energy efficiency, 303
India, 11, 18, 32, 99–101, 103, 116–118, 134–135, 151, 160, 190–192, 195–197, 202, 273, 281, 290, 313
Indian Monsoon, 33
Indian Ocean Dipole, 165, 171, 176, 178, 182, 301, 312, 318
Indonesia, 15, 33, 53, 123, 133, 173, 194, 198, 229–230
integrated phase classification, 189
Intergovernmental Panel on Climate Change, 74–75, 81–82, 110, 268, 283, 300, 305, 309
internal variations, 168
Isaac Newton, 26, 44, 48, 73, 185

Jacob Bjerknes, 193
Japan, 11, 99, 135, 137, 152, 160
Jet Streams, 31
John Snow, 108
Judea Pearl, 154

Kangaroo Island, 252
keeling curve, 30
Kenya, 2, 4–5, 15, 123, 164, 173–174, 177, 214, 226, 232
Kepler, 26
Kerala, 11, 134
koalas, 252
Koninkliijk Nederlands Meteorologisch Instituut, 137
Korea, 160, 191

La Niña, 13–14, 18, 176, 180, 187, 191, 193, 212–216, 223–224, 226, 228, 232, 274, 301, 313, 318
latent heat, 56
longwave radiation, 48

Madagascar, 133, 160
Mali, 103
Mars, 30, 74
methane, 38, 66, 70, 270, 276, 303
Mexico, 33, 187, 203, 284
Milky Way, 27, 35, 37
moral consequences, 182
moral culpability, 113
Mozambique, 188, 199
Munich Re, 7–9, 151, 160–161, 241, 310, 320

National Academies report, 148, 311
Natural Catastrophe Database, 7–8
nebulae, 26
negentropic systems, 57
negentropy, 45–46, 58
Nepal, 135, 151, 273
new normal, 79, 100
Nicolas Léonard Sadi Carnot, 43
Niger, 103
Nigeria, 103, 135

ozone, 28, 56, 66

Pakistan, 99–100, 103, 116–117, 135, 198, 273, 303
Palmer Drought Severity Index, 238–239, 244
Papua New Guinea, 123
pastoralists, 2, 127, 219
Pearl Causality, 155
Peru, 33, 135, 193–194
Phase 5 Coupled Model Intercomparison Project, 84, 138, 227
phase transition, 144
Philippines, 133, 135, 191

photosynthesis, 35, 38, 58, 125, 235, 275
population growth, 75, 105, 114, 116, 268–269, 271, 295, 301
positive feedback, 55, 144, 267, 276
poverty, 1–2, 4, 11, 19–20, 133, 191, 296, 299
prediction, 11, 15, 18, 63, 85, 107, 165, 169, 180, 185, 193, 211, 215, 233, 301
preindustrial simulations, 207
pressure fields, 222
pressure waves, 27

radiative balance, 47, 110
radiative forcing, 110–111, 270, 272
radiative transfer, 29, 72, 75
relative humidity, 61, 128, 130, 234, 240, 256
Representative Concentration Pathway, 104, 110, 261, 268, 278
ribonucleic acid, 57
Ridiculously Resilient Ridge, 240
Royal Society for the Prevention of Cruelty to Animals, 252
Rudolf Julius Emanuel Clausius, 44

saturation vapor pressure, 256
sea surface temperature gradient, 170, 172, 179–180, 222, 315
sea surface temperature gradients, 167–168, 175
shortwave radiation, 48
Sierra Leone, 10
Sir Clive Granger, 154
Sir Gilbert Walker, 32, 193–194, 313
societal change, 183
Somalia, 2, 15, 164, 173–174, 177, 213–215, 218–221, 225–226, 232, 288

South Africa, 188, 191, 201
South Sudan, 103, 164, 197
Southern Oscillation, 32, 190, 193, 307
Spaceship Earth, 38, 294
star generation, 27
stomata, 35, 235–236, 275
storyline, 150, 153, 311
stratosphere, 28
strong force, 26, 28
Sudan, 103, 191, 202, 302

Tahiti, 33, 193
Tasmania, 99
temperature inversion, 55
thermal energy, 107
thermal infrared longwave radiation, 51
thermodynamic control, 130
thermodynamic efficiency, 43
Thomas Fire, 60, 87, 91
thought experiment, 128, 170
trade winds, 31, 33–34
transpiration, 14, 26
troposphere, 31, 33, 47, 56, 144, 223
Turkey, 103
Type I errors, 149
Type II errors, 149

Uganda, 103, 164, 174, 177
United Nations Office for Disaster Risk Reduction, 133, 310

Van Allen belts, 28, 38
vapor pressure deficits, 256, 263, 265–267, 318
Venus, 30, 74
vertical velocity, 128, 130
Vietnam, 160, 191
vulnerability, 7–8, 92–93, 101, 113

Walker Circulation, 32–33, 53–54, 194–195, 225, 228
Warm Pool, 33, 214
Western V, 226, 230–231
wildfire extent, 19, 237–238, 266, 301

World Meteorological Organization, 148, 311

Yuval Harari, 287

Zimbabwe, 15, 188, 199, 201